TRANSACTIONS OF THE

AMERICAN PHILOSOPHICAL SOCIETY

HELD AT PHILADELPHIA

FOR PROMOTING USEFUL KNOWLEDGE

VOLUME 67, PART 5 · 1977

Spanish Red

An Ethnogeographical Study of Cochineal and the Opuntia Cactus

R. A. DONKIN

FELLOW OF JESUS COLLEGE, CAMBRIDGE

THE AMERICAN PHILOSOPHICAL SOCIETY

INDEPENDENCE SQUARE: PHILADELPHIA

September, 1977

Library of Congress Catalog
Card Number 77–76426
International Standard Book Number 0–87169–675–4
US ISSN 0065–9746

PREFACE

Cochineal was the most widely traded and, next to gold and silver, the most valuable product of the Spanish Indies. Moreover, in no other colonial enterprise were the aboriginal associations so pronounced; production was left almost entirely in the hands of the Indian population. An embargo on the export of live insects was maintained for over 250 years; the bulk of the dyestuff, however, was shipped to Spain. Cochineal must often have been among the first New World commodities to appear in the markets of Europe and Asia, where it added to the range of reds and to some extent displaced established pigments, some also of insect origin.

Evidence for the use of cochineal in the highlands of Middle and South America before the Spanish conquest is conclusive, if hardly abundant. From the beginning of the sixteenth century, information on the rearing of the insect and the preparation of the dyestuff may be found in *codices* of native origin, natural histories, official reports and historico-geographical surveys. For the Old World, the sources include herbals, floras, and various scientific reports; state papers; trading records and commercial directories; travelers' accounts; and, from about 1820, pamphlets written to advise prospective farmers. The present attempt to map the extent of cochineal culture (in the New World) and the introduction of insects and host cacti (in the Old) must necessarily be regarded as provisional.

SPANISH RED

An Ethnogeographical Study of Cochineal and the Opuntia Cactus

R. A. Donkin

CONTENTS

INTRODUCTION

COLORS, DYES, AND HUMAN CULTURE

Colors are an integral part of human culture, and they have been the subject of philosophical enquiry from the time of Aristotle. The use of pigments, chiefly red and blue, is known from the late Palaeolithic in the Old World, and there is no ethnographic record of a people without a sense of color. Powers of discrimination appear to correspond with cultural level generally. Colors may be esthetically pleasing, emotionally disturbing, and socially symbolic. They are often associated with particular moral or spiritual qualities,[1] with objects of religious veneration,[2] and with rank and position or caste.[3] In many parts of the world, colors are or have been used to represent the four traditional elements and the four cardinal points. Among the ancient Mexicans, red stood for east (or west), white for west (or east), blue for south, and black for north; four kinds of rain, controlled by the god Tlaloc, were also given color emblems.[4] The contemporary Maya identify the cardinal points with different combinations of black, white, yellow, and red.[5] Almost everywhere, red as the color of fire, the sun, and blood (and thereby life itself) has unusual significance, symbolizing magnanimity and fortitude, majesty and power, both temporal and supernatural.[6] (fig. 1) "Surrounding accompaniments [of red]," observed Goethe, "always have a grave and magnificent effect." [7]

> Contrary to the other chromatic colours, red has its special name in all current languages . . . . [and] the term for red usually proves to be one of the oldest colour terms of a given language. Also, unlike the other chromatic colours, red as a rule is indicated by a native term, and not by one borrowed from other languages.[8]

In Náhuatl, *tlapalli* denoted "color" in general and "red" in particular (similarly in Spanish, *color* and *colorado*). The hieroglyph for the village of Tlapan (eastern Guerrero) was "a circle with the interior painted red . . . . and containing the mark of a human foot-print." [9] Thus to give colors phonetic value was very unusual, if not confined to Mexico.

Pigments and dyes have been used as body and facial paint; on textiles and artifacts of all kinds; and in food and medicinal preparations. Among the many inorganic materials or earths, oxides of copper and iron

[1] Whitney, 1586: pp. 134–135; Chevalier, 1937: p. 31.

[2] Fickeler, 1966: pp. 98–103.

[3] Turner, 1937: p. 57. *Eranos* **41**, 1972, brings together important papers on color symbolism and imagery. See also Hermann, 1969: cols. 369–372, 381–382, 408–409, 431–432.

[4] Westheim, 1948: p. 2579.

[5] Charencey, 1879: pp. 5–69; Roys, 1933: p. 170; Villa Rojas, 1969*a*: p. 241, 1969*b*: p. 271; Laughlin, 1969: p. 175.

[6] Beyer (1921), 1965: pp. 482–487 (ancient Mexicans); Wisdom, 1940: p. 430 (Chorti, Guatemala); Reichel-Dolmatoff, 1968: p. 122 (Tukano, Amazonia); Reina, 1969: p. 129 (Pokomam and Chorti, Guatemala). In the Chinese empire, mandarins of the first grade wore a red stone button on their caps, those of the second grade a red coral button (there were nine grades in all). In Japan in the tenth century, garments of "deep purple and deep scarlet" were associated with high office (Reischauer, 1937: **1**: p. 311); likewise in Black Africa (Zahan, 1972: p. 373). Throughout the Arab world, "the use of red paper in official correspondence was a prerogative of exalted rank and a sign of enjoying special favour" (Martin, 1912: **1**: p. 106). According to the *Chronicle of the Carmelites in Persia,* Shah 'Abbas I (*ca.* 1587–1629) wore "red silk, with a scarlet tunic" on festival days (*Anon.,* 1939: **1**: p. 285 [1608]). In Annam *ca.* 1825 the use of red ink was reserved for the sovereign (Tsaï-tin-lang, 1878: p. 85).

[7] Goethe (1810), 1840: p. 315.

[8] Kouwer, 1949: p. 14.

[9] Brinton, 1890: p. 224.

In colores.
To EDWARDE PASTON *Esquier.*

THE dier, loe, in smoke, and heate doth toile,
Mennes fickle mindes to please, with sundrie hues:
And though hee learne newe collours still to boile,
Yet varijng men, woulde faine some newer choose:
And seeke for that, which arte can not deuise,
When that the ould, mighte verie well suffise.
And some of them, here brieflie to recite,
And to declare, with whome they best agree:
For mourners, *blacke.* for the religious, *White.*
Which is a signe, of conscience pure, and free.
The *greene*, agrees with them in hope that liue:
And eeke to youthe, this colour wee do giue.
The *yelowe* next, vnto the couetous wighte.
And vnto those, whome ielousie doth fret.
The man refus'd, in *Taunye* doth delite.
The collour *Redde*, let martiall captaines get.
And little boies, whome shamefastnes did grace,
The Romaines deck'd, in *Scarlet* like their face.
The marriners, the *Blewe* becometh well.
Bicause it showes the colour of the sea:
And Prophettes, that of thinges deuine foretell,
The men content, like *Violet* arraie.
And laste, the poore and meaner sorte prouide,
The *medley*, *graye*, and *russet*, neuer dy'de.

Ouid de Trist.
Infœlix habitum temporis huius habe
Nec te purpureo velent vaccinia succo:
Non est conueniens luctibus ille color.

Alciat.
Nos sperare docet viridis, spes dicitur esse
In viridi quoties irrita verò cadit.

FIG. 1. Geffrey Whitney, *A Choice of Emblemes,* Leyden, 1586.

(ochre, Náhuatl *tecocahuitl*) were prominent. Hundreds of species of plants have been valued as sources of dye. Some, such as madder and indigo, achieved world-wide importance, but the majority met only local or regional needs. The processes of extraction and preparation varied even when the same plant was employed and were often regarded as secrets or mysteries. Almost every part of the vegetable kingdom has been exploited at some time. In the words of the author of *De Coloribus* (? *ca.* 300 B.C.), "many things are coloured by the flowers of plants, many by the roots, many again by bark or wood or leaves or fruit . . . ." [10]

There were far fewer animal dyes, in fact only two main kinds, prepared from similar species of shellfish and of insect. The dyes obtained, shades of red or purple, were highly concentrated and notably successful on fibers of animal origin. In short, they were the finest natural dyestuffs, rich in tone, generally fast to light and water, and invariably costly. Shellfish dyeing had to be conducted near to the coast; the dried insect products, on the other hand, could easily be exported. Trade in cochineal eventually encircled the globe. In the Old World, knowledge of the nature of the medium remained confused until the eighteenth century, and attempts to introduce the insect itself were unsuccessful for one reason or other until the end of the century, that is to within about sixty years of the production of aniline dyes and the rapid replacement of almost all natural dyestuffs.

[10] Aristotle [School of], 1963: p. 19.

PRINCIPAL RED DYES, OTHER THAN COCHINEAL

Reds obtained from madder, brazil, kermes, Polish cochineal, and lac dominated the dye markets of Europe and western Asia at the beginning of the sixteenth century.[11] The arrival of American cochineal had important consequences, for the new product was soon recognized as a very powerful coloring agent. Another red dyestuff, *annatto* (*Bixa orellana*), of vegetable origin, was also imported from the middle of the sixteenth century, but, being dull in tone and fugitive to light, it was always prized far below cochineal.

VEGETABLE DYES

Madder

Rubia: Members of the Rubiaceae or madder family were widely employed in the preparation of red dyes (color principle alizarin [12]) in both the Old and the New World.[13] Alphonse de Candolle was apparently correct in identifying western temperate Asia and southeastern Europe as "the natural, ancient and undoubted habitation" of the most important species, *Rubia tinctorum.*[14] N. I. Vavilov assigned the cultivated forms to no less than three of his chief centers of origin (India, the Near East, and the Mediterranean lands).[15] Use of the root is reported from Pharaonic Egypt and the Indus valley (Mohenjo Daro) but not, apparently, from Mesopotamia.[16] The Greeks (*erythrodanon*) and Romans (*rubia*) employed *Rubia tinctorum* in the preparation of dyes and medicinal compounds.[17] Dioscorides in his *Materia Medica*

[11] For other "reds" of vegetable origin, see Mairet (1916), 1938: pp. 11–22.

[12] Arabic *lizari* or *alizari* (madder root), probably from *al-'aṣārah,* the juice. Alizarin was first isolated in 1826 (Perkin and Everest, 1918: p. 5).

[13] Schaefer (1941) has a sketch map of the world distribution of *Rubia* and of madder dyeing. One of the leading blue dyes, obtained from species of *Indigofera* (*tinctorum* [India], *suffruticosa* [*America*] and *argentea* [*Africa*]), was also known in both hemispheres.

[14] Candolle (1882), 1964: p. 42.

[15] Vavilov (*ca.* 1935), 1951: p. 28.

[16] Forbes, 1954: p. 246; 1964: p. 107.

[17] Herodotus (fifth century B.C.), 1963–1969: **2**: p. 393; Theophrastus (372–287 B.C.) 1961–1968: **2**: p. 285; Vitruvius (first century B.C.), 1962: **2**: p. 127 (*rubia radice*); Strabo (?63 B.C.–A.D. 24), 1960–1969: **2**: p. 107, **6**: p. 189; Pliny (A.D. 23–79), 1961–1968: **5**: p. 451; **7**: p. 69; Dioscorides (first century A.D.), 1952–1959: **3**: p. 366.

distinguished between wild and cultivated varieties. The plant spread through southern Europe and, presumably later, in areas north of the Alps. Holland was one of the principal areas of cultivation at the close of the Middle Ages. *Rubia peregrina,* also a native of the Near East, was cultivated in Turkey and Provence,[18] and *Rubia cordifolia* in India (Hindi, *majīth*), Java and China.[19] Silks of the Han period (206 B.C.–A.D. 220) were dyed with madder.[20]

Relbunium: Species of *Relbunium* are distributed through the tropical and temperate regions of America from Mexico to Argentina.[21] Roots and leaves were found at the necropolis of Ancón on the north coast of Peru;[22] and G. A. Fester has reported *Relbunium* (? *nitidum*) in colored textiles from Paracas and from prehistoric tombs in northern Argentina.[23] A *relación* of Santa Cruz de la Sierra (eastern Bolivia) in 1571 mentions "colorado de raíces," probably *Relbunium*.[24] The plant known in Quechua as *chchapi* or *chapichapi,* and described as a source of red dye by Bernabé Cobo (1653),[25] has been identified as *R. microphyllum*. *R. hypocarpium* was used in southern Colombia (Pasto, Coaiquer, and Quillacingo peoples), Peru and northern Chile (Araucanians).[26] *Relbunium* sp. apparently was not exploited in Middle and Central America, its place being taken by *Bixa orellana* and cochineal.

Bixa orellana

Bixa (*achiote, urucu, annatto*) was put to a variety of uses, in both Middle and South America, from the pre-Columbian era to the advent of aniline products.[27]

Bixa orellana, an evergreen shrub, belongs to the frost-free and humid tropics, but the dried seeds from which the dyestuff is prepared were traded throughout the highlands. The plant was introduced to the Old World soon after the Spanish conquest of Mexico and it now has a pan-tropical distribution.

Dyewoods

Brazil-wood (chiefly *Caesalpinia echinata*) seems to have been locally exploited from remote times in the New World.[28] Europe obtained brazil (*C. sappan*) from India and the East Indies in the Middle Ages.[29] Chaucer (Controller of the English Customs, 1374–1386) referred to it, and to "grain" or kermes, in the epilogue to the Nun's Priest's Tale:

> Him nedeth nat his colour for to dyen
> With brasile, ne with greyne of Portyngale

From the middle of the sixteenth century the tropic coasts of America became the main areas of supply.[30]

ANIMAL DYES

Shellfish

In both the Old and the New World, the rarest and costliest dyestuff was obtained from shellfish. Several species secrete an almost colorless fluid which has the effect of staining fabrics, first yellow-green and then, after exposure to sunlight, red or violet or deep purple according to the species and the processes employed.

Old World: The subject of "Tyrean purple" has been extensively studied.[31] The main area of production was the eastern Mediterranean, above all the shores of the Levant, but dye factories were also started in Phoenician and later times in Greece, Crete, Cyprus, Sicily, Dalmatia, Asia Minor, Italy, Sardinia, southern France, Spain, Libya, and even as far afield as the shores of the Red Sea and the Atlantic coast of northwest Africa. From time to time, similar but inferior dyes have been extracted from the species *Purpura* [*Thais*] *lapillus* around the coasts of Ireland, Brittany, and Cornwall.[32] The explorer J. Nieuhoff found (1662) that a red dye or "ink" was prepared from shellfish in the islands around Java.[33]

The most important species within the Mediterranean were *Murex trunculus* and *M. brandaris,* and on the Atlantic shores *Purpura* [*Thais*] *haemastroma.* Pliny distinguished between the two (*pelagia* or *purpura* and *buccinum* respectively) and also gives a detailed description of the natural history of the shellfish, the manufacture of the dye, and the symbolic significance of "purple."[34] The earliest evidence is from the middle

18 Beckmann, 1846: p. 113.

19 Henry, 1887: p. 238; Vavilov (*ca.* 1935), 1951: p. 26.

20 Pfister, 1934–1940, 3: pp. 66–67; Willetts, 1958, 2: p. 241.

21 Fester and Lexow, 1943*b*: p. 233; Towle, 1961: p. 89; Girault, 1966: p. 170.

22 Trémeau de Rochebrune, 1879: p. 347 (*Rubia nitida*); Wittmack, 1888: p. 347.

23 Fester and Cruellas, 1934: p. 156; Fester and Lexow, 1943*b*: p. 234; Fester, 1953: pp. 13–16, 1954: pp. 238–240.

24 Jiménez de la Espada (ed.), 1965: 1: p. 408.

25 Cobo, 1956: 1: p. 189.

26 Hernández de Alba, 1947: p. 943; O'Neale, 1949: p. 124. Concerning Chile, see Frezier, 1717: p. 78 (*reilbon*), and J. I. Molina (1787), 1809: p. 115 (*relbun: Rubia chilensis*).

27 Donkin, 1974: pp. 33–56.

28 La Cruz (1552), 1939: p. 133 (*huitz-quahuitl*); Sahagún (*ca.* 1570), 1963: p. 241 (*uitzquauitl*); Wittmack, 1888: p. 347; Noyes, 1932: p. 313 (*C. platyoba*); Roys, 1943: p. 51; O'Neale, 1949: p. 124; Friedberg, 1958: p. 419; Osborne, 1965: pp. 39–40 (*Haematoxylon brasiletto, Bocconia arborea*). Fester (1940: pp. 234–235) observed that none of the chroniclers of Peru mentions "palo de Brazil" but that it appears in a document of 1703. There is a Quechua word, *tara,* for *Caesalpinia pectinata* (Cook, 1925: p. 100).

29 Leix, 1937: p. 21 (1321); Leggett, 1944: p. 50 (1193); Magoun, 1955: p. 121 (*ca.* 1100).

30 Chevalier, 1943: p. 329.

31 Commencing perhaps with Fabius Columna, 1616. See Dalman, 1937: pp. 78–80, and Jensen, 1963: pp. 104–118.

32 Cole, 1689; J. W. Jackson, 1916: pp. 15–17; T., 1937: pp. 131–132.

33 Nieuhoff, 1704: p. 353. Read (1937: p. 240, based on the *Pen T'sao Kang Mu,* 1596) gives the Chinese name for *Purpura tumulosa.*

34 Pliny, 1961–1968: 3: pp. 247–257. See also Aristotle (*Historia Animalium*) 1968–1970, 2: pp. 147–149; Vitruvius, 1962: 2: p. 127; Strabo, 1960–1969: 7: p. 269.

of the second millennium B.C. (Phoenicia and Crete). Homer refers to the "purple" or "scarlet" of Caria and Lydia (Asia Minor),[35] and the prophet Ezekiel (sixth century B.C.) to purple cloth from the isles of Elishah or Laconia (Peloponnesus).[36] Hellenic and Roman sources contain scores of descriptions and allusions, and evidence for the whole of antiquity is preserved in innumerable shell middens. After the fall of Rome, the industry survived chiefly at places around the sea margins of the Eastern Empire. Purple was then used for dyeing silk and in the production of sacred books or codices (grounding the parchment before illustration or lettering in gold). The expansion of Islam, particularly the Turkish invasions, had a disastrous effect on output of the dye, and it was virtually unknown at the time of the discovery of the New World.

New World: The shellfish-dye industry of the Mediterranean had a close, but historically independent counterpart on the west coast of the Americas, notably between Michoacán (Mexico) and the Gulf of Guayaquil (Ecuador) (map 3).[37] *Purpura patula* was exploited in Mexico and Central America, *Thais kiosquiformis* in northwestern South America. The former species also occurs in the Caribbean but was valued only as food. In contrast with the various Old World shellfish, *P. patula* secretes the dye fluid freely; destroying the creature is unnecessary and it can be returned to the sea and "milked" at suitable intervals.

The question of the use of shellfish dye in pre-Columbian times has evoked discussion since E. Von Martens's paper "Purpur-Färberei in Central-Amerika" was published in 1898. The evidence, while not large, is fairly conclusive. Cotton fabric recovered from graves in the Ica valley, Peru (? first century B.C.), shows purple markings, although exactly where the dye was obtained has not been determined.[38] Fragments of textile (*ca.* A.D. 1500) from Chiapas, Mexico, had been similarly treated.[39] Purple paint was employed in some of the Mixtec and Zapotec codices;[40] particular deities and items of clothing of exalted persons, both male and female, were thus colored. This and the symbolic significance of purple in the Old World derive from the strong esthetic and emotional appeal of a dye that was also necessarily in short supply.

Fernández de Oviedo y Valdés (1535) was the first Spaniard to remark upon the American shellfish—"conchas u ostras de la púrpura en el golfo de Orotiña" (Nicoya, Costa Rica).[41] Concerning the same area, the English Dominican Thomas Gage wrote (1648) of "certain shells wherewith they make this purple dye . . . [to] colour a kind of thread called *pita,* which is a very rich commodity in Spain."[42] However the bulk of the dyestuff has always been used in Indian craftwork; and the amount of dyed thread (*hilo morado*) or cloth sent to Spain was apparently very small. In 1704, Father Alcazar published an account of "un petit limaçon des Indes (Nicaragua) qui paroît être le Murex des Anciens."[43] About forty years later, Antonio de Ulloa described the industry of the coast of Ecuador, as well as of Nicoya, thus:

> On the coast belonging to this lieutenancy (Santa Elena) is found that exquisite purple, so highly esteemed among the ancients; but the fish from which it is taken, having been either unknown or forgotten, many moderns have imagined the species to be extinct. This colour however is found in the species of shellfish, growing on the rocks washed by the sea. . . . The jurisdiction of Nicoya in the province of Guatemala also affords the species of turbines, the juice of which is used in dyeing cotton threads and, in several parts, for ribbands, laces and other ornaments.[44]

The industry extended north and south of Nicoya into, respectively, the jurisdictions of Nicaragua and Veragua.[45] Altogether several Indian groups—Quepo, Boruca, Guaymí, and Orotiña—were involved.

The third important area lay along the coast of the Mixteca Baja and of Oaxaca, Mexico, where the shellfish was known as *caracól* or *coacoyul.* Francisco de Burgoa in 1670 referred to the "caracolillos de púrpura" of the Huave and Chontal of southern Oaxaca.[46] The word *caracolillo* was still in use in 1909[47] and, more strangely, was reported by Ulloa from the vicinity of Santa Elena.

Cotton thread was dyed by the seashore, then marketed sometimes far inland. From the middle of the nineteenth century dyeing has been an increasingly rare occupation, confined to stretches of isolated coast,[48] and hardly found at all in South America.

Insects

Insects have been exploited by man in his search for dyestuffs over several millennia. They have also been important as a source of food (proteins, fats, sugar)

[35] *Iliad,* 1950: p. 80, 1967–1971: 1: p. 163.

[36] Ezekiel 27: 7, 16.

[37] After Gerhard (1964), the most important recent account.

[38] Gerhard, 1964: p. 179. Saltzman, Keay and Christensen (1963: pp. 244–245) also reported the presence of shellfish purple on samples of fiber (400 B.C.–A.D. 100) from Peru.

[39] Johnson, 1954: pp. 142–145; Gerhard, 1964: p. 179.

[40] Nuttall, 1909: p. 381.

[41] Oviedo y Valdés, 1959: 4: p. 424. There is no description in Sahagún *Codex Florentino, ca.* 1570, or in Hernández *Historia Natural,* 1571–1576.

[42] Gage, 1928: p. 358.

[43] Alcazar, 1704: pp. 1772–1774.

[44] Ulloa (1748), 1760: 1: p. 176. See also Jinesta, 1940: pp. 303–304.

[45] Squier, 1852: 1: p. 286. Dézallier d'Argenville (1742: p. 181) called attention to the shellfish dye of "Panama" (*pourpre de Panama*) and of "Guatemala."

[46] Burgoa, 1934: 2: p. 406.

[47] Nuttall, 1909: p. 369.

[48] In addition to the authorities noted above, see Murillo, 1925: pp. 22–23; Barrett, 1925: p. 258 (Esmeraldas, Ecuador); O'Neale, 1949: p. 125 (El Salvador); D. Stone, 1949: pp. 16–17 (Costa Rica); Brand, 1960: pp. 15, 29 (Michoacán, Mexico); Nader, 1969: p. 341 (Oaxaca, Mexico).

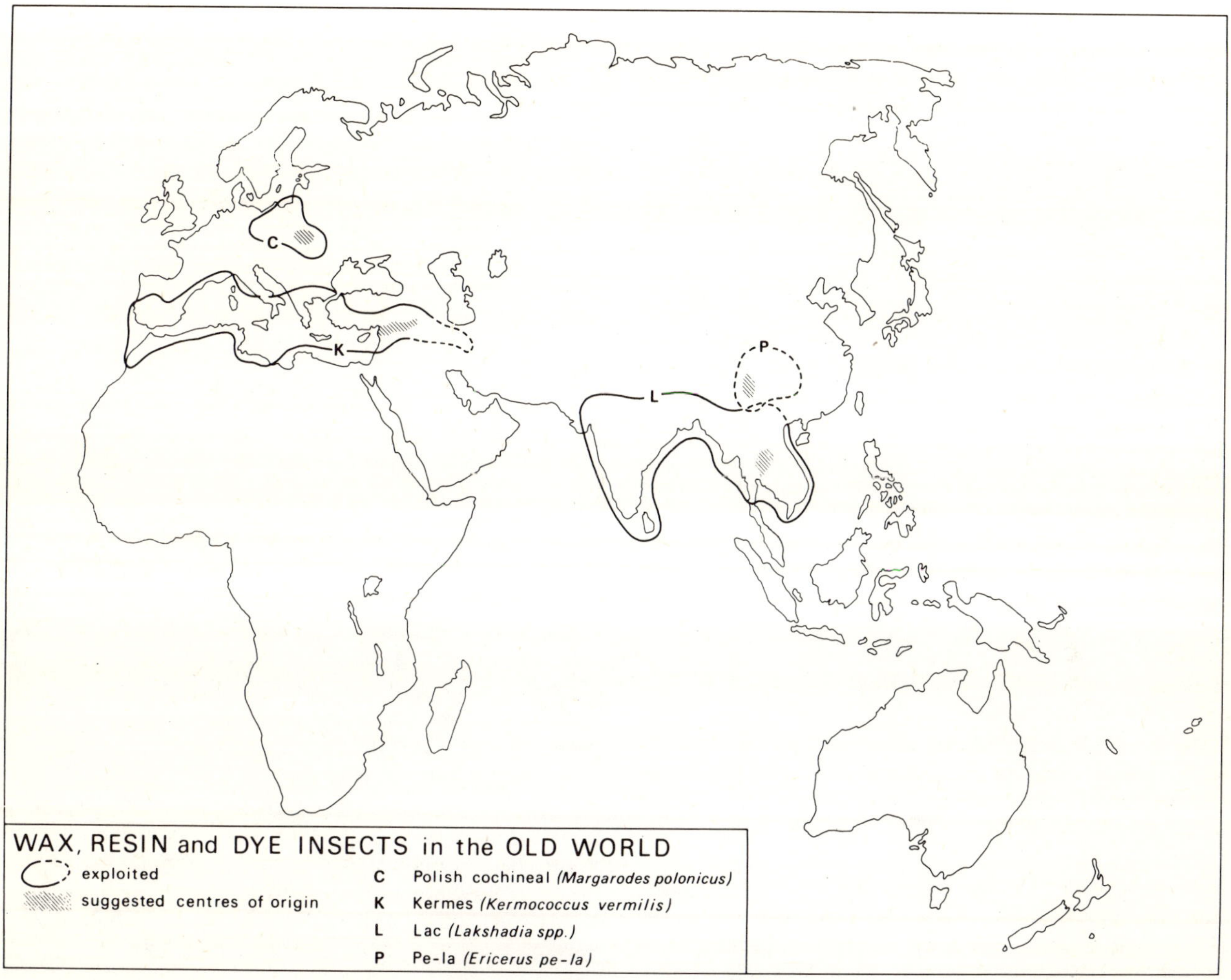

MAP 1. Wax, resin and dye insects in the Old World.

and drugs,[49] of wax and resin, and in supplying a luxury fiber, silk. Exploitation of the wax-resin-dye insects (all members of the superfamily Coccoidea) belongs, in the New World, chiefly to the highlands of Middle and South America, and in the Old World to a broad and possibly continuous zone stretching from the shores of the Mediterranean to western China. Whether the different forms of exploitation are historically connected is at present unclear, but nevertheless likely. In the case of the lac insect, resin and dye are complementary products. Everywhere the work of collecting the respective insects and preparing the product is time-consuming and also calls for certain broadly related skills. Domestication proceeded through human management of the host species (improvement, selection) and by facilitating successful breeding.

Kermes: Kermes was the most important of the animal dyes of Old World provenance. The insect (*Kermococcus vermilis*) breeds on the kermes oak (*Quercus coccifera*) which is found around the shores of the Mediterranean and in parts of the Near East. Extraction of the dye appears to have commenced in lands to the east of the Mediterranean (map 1) where the true nature of the product was understood at a very early date.[50] Among the Greeks and the Romans, for two or three centuries on either side of the beginning of the Christian era, there seems to have been some doubt as to whether kermes was an insect or a vegetable substance. Theophrastus (372–287 B.C.) in his *Enquiry into Plants* states that the kermes oak "bears a kind of scarlet berry." [51] Pliny in one place refers to a "berry" (*coccum,* from the Greek *kókkos*);[52] elsewhere to a "little worm" (*scolecium*).[53] Dioscо-

[49] Hardwick, 1822; Skinner, 1910; Curran, 1937; Bodenheimer, 1947, 1951; Fischer, 1955: pp. 1–37; Ruddle, 1973: pp. 94–101.

[50] Canáls y Martí, 1768; Bodenheimer, 1929: **2**: pp. 188–194; Forbes, 1964: p. 105 and table II.

[51] Theophrastus, 1961–1968: **1**: p. 259.

[52] Pliny, 1961–1968: **3**: p. 259; **6**: p. 294; Strabo, 1960–1969: **6**: p. 189 (*coccus*).

[53] Pliny, 1961–1968: **4**: p. 409.

rides is similarly equivocal.[54] The geographer Pausanius, writing in the second century A.D., came close to the truth: "This shrub the Ionians, as well as the rest of the Greeks, call *kókkos*. . . . There breeds in the fruit of the *kókkos* a small creature . . . . and the blood of the creature serves as a dye for wool." [55]

The antecedents of the word *kermes* are found in several oriental languages: Sanskrit *kirmira, kirmidja,* Pahlavi *karmīr, kalmīr,* Persian *kirmiz,* Kashmiri, *kirmizī,* Armenian *karmir, garmir,* Turkish *qurmuz, kirmiz,* Arabic *qîrmiz,* Hindi *kirmdāna.* The etymological root, *kṛmi* (Skt.), *kirm* (Pahl.), means "worm." The Hebrews, who apparently did not prepare the dye themselves, referred to *karmīl* and also to *tōla'ath shani,* "red worm." [56] Some variant of *kirmiz* was probably first applied to a dye-yielding insect of Armenia and Āẕarbāījān (*Porphyrophora hameli*) [57] and subsequently to *Kermococcus vermilis.* The Phoenicians apparently employed the Mediterranean kermes and one of the earliest references is to a Phoenician merchant in the middle of the second millennium B.C. carrying "rouge extracted from worms" into what is now northern Iraq.[58] The Egyptians obtained kermes—or cloth dyed with kermes—from the Levant and Persia.[59] Production of the dyestuff appears to have spread westward through the Mediterranean lands in Classical times. Pliny tells us that every other installment of tribute from Spain consisted of kermes. It was used as a superior dye for woolens and silk and also in medicinal preparations.[60]

Kermes declined in importance during the Dark Ages but then re-emerged as the most celebrated dye of the Middle Ages when, north of the Alps particularly, it was sometimes known as *vermiculus* (small worm), hence the word vermilion.[61] Kermes-dye was traded through the length and breadth of Europe [62] and also into Persia, India, and China.[63] Brighter than madder and faster than cochineal, it continued to be used for scarlet until the middle of the seventeenth century when it was found that cochineal mordanted with chloride of tin gave a more vivid color. Live insects were never carried to the New World.

Polish cochineal: Polish cochineal [64] is peculiar in that the insect *Margarodes* [*Coccus, Porphyrophora*] *polonicus* lives on the root of the host plant, *knawel* (*Scleranthus perennis*). Both belong to eastern Europe, mainly Prussia, Saxony, Lithuania, Poland, and the Ukraine.[65] Harvesting was laborious work, for the plants had first to be uprooted and then replaced.[66] The insects were killed in the manner of kermes, by emersion in a solution of vinegar or by exposure to vinegar vapor. The industry must date from at least the beginning of the Christian era. The Chinese (? first century A.D.) obtained the product through trade.[67] In eastern Europe during the Middle Ages, Polish cochineal was subject to tithe and it also appears as an item in rents in kind, particularly on monastic estates. This, and the high price of the superior kermes, tended to encourage production. Reports of the marketing of the dyestuff continue into the early nineteenth century.

Lac: The several species of lac insect (*Lakshadia* [*Tachardia, Laccifer*] spp.) are widely distributed in southern and southeastern Asia. They have been reported from many parts of India,[68] as far west as Sind, as well as from Sri Lanka, Nepal, Bhutan, Tibet, Assam, Burma, Siam, Indochina, and Yun-nan (*Lakshadia chinensis* or *siamensis*).[69] The insect is found both wild (*L. communis*) and domesticated, and in association with a large number of host trees.[70] The

[54] Dioscorides, 1952–1959: 3: pp. 403–404.

[55] Pausanius, 1965–1969: 4: p. 585.

[56] Moldenke, 1952: pp. 193–194; Michell, 1955: p. 246; Bodenheimer, 1960: p. 157.

[57] Hamel, 1835: pp. 9–64; Bryce, 1877: p. 227; Laurent, 1919: p. 41; Kurdian, 1941: pp. 105–107; Petrushevsky, 1968: p. 505. The Armenian insect fed on species of grass, *Poa* [*Aeluropus*] sp., *Dactylis* sp. The plain of Ararat is most often mentioned (see González de Clavijo [1403–1406], 1928: p. 143). The dye was traded in western Asia, and some references to "kermes" and "cochineal" from this area are thereby difficult to interpret. The *kirmiz* of al-Iṣṭakhrī [tenth century] (Manandyan, 1965: p. 143), Ibn Ḥawqal [tenth century] (1880: pp. 160–161), the Persian author of "The Regions of the World" [A.D. 982] (Minorsky, 1937: p. 142), and of Ḥamd-Allāh Mustaufī [1340] (1919: pp. 89; *cf.* 1928: p. 45) was presumably the Armenian species.

[58] Bodenheimer, 1960: p. 157. See also Moldenke, 1952: p. 195 (1727 B.C.).

[59] Cotte, 1918: p. 95; Pfister, 1928: p. 242. Ibn Ya 'qūbī (*ca.* A.D. 900) observed that Assiut in Upper Egypt produced "tapis teints au kermès dans le genre des tapis d'Arménie" (Wiet [ed.], 1937: p. 186).

[60] Cotte, 1918: p. 100. *Cf.* Eichstadius, 1635; *Pharmacopoea,* 1659: p. 26; 1682: pp. 67, 181; Bartholinus, 1672; Pomet, 1709: pp. 31–32; Geoffroy, 1736: p. 370; Hellot (1789), 1901: p. 96; Planchon, 1864: pp. 38–47.

[61] The description "vermilion" of mineral substances (e.g. cinnabar) apparently arose from their resemblance to crimson of insect origin. "Vermilion" mixed with ceruse was the basis of a rose-colored pigment used in miniature painting in medieval Persia and India (Martin, 1912: 1: p. 108).

[62] Pegolotti (*ca.* 1310–1340); 1936: pp. 123 ff. (*grana*).

[63] Heyd, 1886: 1: pp. 272, 279; 2: pp. 282, 350, 607–609; Pfister, 1934–1940: 3: p. 67.

[64] So described by Linnaeus, 1772: p. 66, and Linck, 1787: p. 14. *Cf.* Moffett (1634), 1658: p. 1098.

[65] Fernald (1903: p. 29) refers to *M. polonicus* in Mexico.

[66] Bodenheimer (1928–1929: 2: pp. 200–207) has an illustration of a trowel used in harvesting.

[67] Pfister, 1934–1940: 3: pp. 67, 69. On the early use of Polish cochineal in western Asia, see Pfister, 1935: p. 46.

[68] Roxburgh, 1799: pp. 361–364; Watt, 1901: pp. 181–347; Stebbing, 1907: pp. 24, 84; Sleeman, 1939: pp. 47–51.

[69] Mahdihassan, 1948: p. 444. Bühler (1948: p. 2482) repeats a report (by Höltker, unlocated) of a red insect dye (? lac) from New Guinea. Watt (1889–1893: 4: p. 570) includes "some islands of the eastern archipelago" in describing the distribution of the lac insect. Fernald (1903: p. 125) adds British Guiana, without stating when and in what circumstances it was introduced.

[70] Watt, 1889–1893: 2: p. 410; 1901: p. 181. In India

female secretes a resinous substance (stick-lac) from which shellac and lac dye (lac lake, cake lac) are prepared.

The richest dye came from what was probably the earliest center of production—Indochina, Siam and neighboring parts of Burma (Pegu—Martaban), whence it was exported to countries to the east and west.[71] The word *lac* seems to be of Cambodian origin (*le-ak*), later Indianized (Sanskrit *lâkshâ*).[72] A Chinese work of the early fourth century A.D. refers to the lac of Indochina, used for dyeing raw silk.[73] A piece of the resin, apparently from Siam and dating from the eighth century, is known from Japan.[74] In the thirteenth century lac-dye was one of the products of Tongking (*Kiah-chï*) and of Cambodia (*Tchin-la*).[75]

Small amounts of dye and of shellac probably began to reach the eastern Mediterranean from India around the beginning of the Christian era.[76] Several references to the dye have been found in Arabic sources.[77] In medieval times *lacca,* along with various dyewoods, was imported through Venice.[78] Perhaps the earliest first-hand report of the insect by a European—Garcia da Orta, a Portuguese physician—was printed in Goa in 1563.[79] In India *lakh* was employed to dye or stain leather, wool, and silk, and also as a substitute for henna (powdered leaves of *Lawsonia inermis*). It is particularly successful on silk, which explains the early Chinese interest. In Europe the insect product was sometimes mixed with madder.[80]

The white-wax insect of western China, *Ericerus pe-la* (la-tchong), is said to have been exploited from the thirteenth century;[81] reports by Europeans commence soon after 1600.[82] The wax (replacing beeswax) was used to coat tallow candles and pharmaceutical preparations. Production centered upon Szech'uan and northern Yun-nan, in fairly close proximity to districts producing lac. The two activities may be historically related. The Emperor K'ang-hsi (1662–1722) alluded to *pe-la* in discussing the origin of, apparently, lac dye.[83] Like *Lakshadia* sp. the white-wax insect lives on a variety of host trees. A. Hosie (1890) described how the creatures were propagated on tree privet (*Ligustrum lucidum*) in the valley of An-ning and then carried a distance of 200 miles north east, to the neighborhood of Chia-ting (Lo-shan), there to deposit wax on the branches of ash (*Fraxinus chinensis*).[84] By the close of the nineteenth century the industry was already in decline; demand (following the introduction of kerosene), production, and prices had all fallen sharply over a period of twenty years.[85] In 1939 about 2,500 tons were exported.[86]

1. COCHINEAL

HOST

Cochineal insects are parasites of cacti belonging to two closely related genera, *Opuntia* and *Nopalea;* the latter includes eight or nine species, the former over 200.[1] There are also considerable intra-specific variations in spininess and in the color and size of the joints (*pencas,* botanically phylloclads), flowers, and fruit. The domestication of certain species has tended to multiply such differences.

The family Cactaceae, with the possible exception of the small genus *Rhipsalis,* belongs to the New World. Opuntias and nopaleas in particular have been exploited and to a varying extent cultivated since pre-Columbian times. Spiny species make very effective enclosures,[2] and the spines themselves were used as needles. The early Spanish historians, commencing with Oviedo y Valdés (1526), compared the fruit (*tuna*) to large figs,[3] hence the subsequent description "Indian fig" (*higuera de las Indias*). Toribio de Motolinía

chiefly *Butea frondosa* (palas or dhak tree, Sanskrit *lákshá-táru*), *Ficus religiosa* and *Schleichera trijuga* (kusum).

[71] Linschoten, 1598: pp. 116–117; Meilink-Roelofsz, 1962: pp. 65, 70.

[72] Mahdihassan, 1954: p. 321. *Cf.* Laufer, 1919: pp. 475–478.

[73] Mahdihassan, 1950*b*: pp. 289–290 (the *Wu Lu* of Chang Po); Laufer, 1919: pp. 475–476; Schafer, 1957: p. 135. Lac dye may have been used as early as the Han dynasty, 206 B.C.–A.D. 220 (Willetts, 1958: 2: p. 241 n. 3). See also Pfister, 1934–1940: **3**: pp. 64–65, 67, 69; and K'ang-hsi (*ca.* 1700), 1779: p. 478, referring to the T'ang period (618–906).

[74] Mahdihassan, 1952*b*: pp. 39–43.

[75] Hirth and Rookhill (trans. of *Chu-fan-chi* by Chao Ju-kua), 1911: p. 46; K'ang-hsi (*ca.* 1700), 1779: p. 478.

[76] Schoff (ed. *The Periplus of the Erythraean Sea*), 1912: p. 24; Warmington, 1928: p. 178; Pfister, 1935: p. 14 n. 57. *Cf.* O'Conor, 1876: p. 8.

[77] Forbes, 1964: p. 107. See also Goitein, 1973. According to Mahdihassan (1954: p. 322), the Arabic names derive from Indochina and Java rather than India.

[78] Pegolotti (*ca.* 1310–1340), 1936: p. 137, 141, 144. The question is complicated by the fact that *lacca* or *lakka* could also refer to dyewood (Mahdihassan, 1954: p. 321)—? *laka-wood.*

[79] Orta, 1913: pp. 240–250 (*lacre*).

[80] In England about 1830, the use of lac dye was increasing and the amount imported (? 600,000 lbs) was approximately twice that of cochineal (McCulloch, 1832: p. 695).

[81] Hedde and Rondot, 1849: p. 56; Julien and Champion, 1869: pp. 106–114; R. Blanchard, 1883: pp. 29–32; Wilson, 1913: 2: p. 100.

[82] Trigault (1615), 1618: p. 23.

[83] K'ang-hsi, 1779: p. 478 ("le [rouge] Tsée-y est tiré d'un insecte (*Tsée-pien*) qui ressemble à celui de l'arbre qui donne la cire").

[84] Hosie, 1890: pp. 189–201. There is a very similar account in Reclus, 1882: p. 417. See also Cooper, 1871: p. 429.

[85] MacGowan, 1850: pp. 93–95; Silliman, 1871: p. 684.

[86] Chiao and Pen, 1940: p. 108.

[1] Bravo, 1937: pp. 105–111 (*Nopalea*), 111–202 (*Opuntia*); Backeberg, 1958: 1: pp. 389–628 (*Opuntia*), 628–633 (*Nopalea*); Britton and Rose (1937), 1963: 1: pp. 33–39, 216 (*Nopalea*), 42–215, 216–225 (*Opuntia*).

[2] Early remarked in Lewin (ed.) (*ca.* 1600), 1958: p. 49; *Anon.*, 1693: p. 503. In parts of Brazil, the opuntia is known as *tababeira* (*taba,* a Tupi word for village; *beira,* border or edge). See also Wavrin, 1937: p. 60.

[3] Oviedo y Valdés, 1950: p. 240.

(1541) observed that there were *muchas especies* of nopal and that the fruit was appreciated by the Spaniards as well as by the native inhabitants of the Indies.[4] Sigvald Linné (1948) concluded that it was the most important native fruit of the valley of Mexico and probably the only one cultivated in pre-Columbian times[5] (as to the latter, he was certainly incorrect). The preparation of a kind of preserve (*miel, melcocha, queso de tuna*) by evaporating the juice of the *tuna* is probably of ancient origin;[6] and so too is the custom of fermenting the juice to make the beverage *colonche* (Náhuatl, *nochoctli*). Carl Sartorius (1850) reported that *colonche* was consumed "as a substitute for wine at the festivals of shepherds and mountaineers."[7]

In Mexico, the joints of the plant, cooked with chili (*Capsicum* sp.), were recommended as a cooling diet,[8] and the sap (*zumo*) was extracted where drinking water was scarce.[9] Young and tender *pencas* are still consumed, and the flowers are sometimes added to stews.

Products of the opuntia were also employed in folk medicine. They occupied an important place in the pharmacopoeia of the Callahuayas, traditional doctors of the Andes.[10] Within the Aztec realm, the sap or pulp, mixed with other substances, was used for treating burns,[11] and in Hispaniola, according to Oviedo, the pulped *pencas* were applied to broken limbs.[12] Later, the medicinal and therapeutic properties of the exotic plant appear to have been widely advocated in the Old World.[13]

The species most commonly used in rearing cochineal was *Opuntia Ficus-indica*,[14] found today only in cultivation. The somewhat misleadingly named *Nopalea cochenillifera* (Salm-Dyck, 1849) occupied, at best, second place.[15] The Náhautl name for the plants of both genera was *nopalli* or *nochpalli*, and for the fruit *nochltí* (cultivated, *tlatocnochtlí*).[16] The Spaniards referred to them as *tunal* and *tuna* respectively,[17] having already adopted the Carib word *tun* (fruit or seed).[18] The Mexicans called cochineal *nocheztli*,[19] that is "blood (*eztli*) of the *nochtlí*." Perhaps this refers to a time when only the red fruit (*tuna pintadera, tlapalnochtlí*) of certain opuntias was used in dyeing.

Nochtlí or *nocheztli* is an element in the Náhuatl place-names Nochixtlán (*tlán*, place), Nochixtongo, Nochtepec, and Nochcoc. In the Codices Mendoza (1540's) and Yanhuitlán (1545–1550) the glyph or symbol for these townships is a nopal or, in the case of Nochixtlán, two nopal joints and a bowl of red pellets.[20] Nochixtlán lay in the heart of the Mixteca, perhaps the earliest center of cochineal production. The Mixtec knew the place as Nunduco (*nuhu*, land; *n'duco*, cochineal) or Nuataco (*ataco*, insects).[21] The name of the capital of the federal empire of Mexica-Culua, Tenochtitlán, also apparently includes the element *nochtlí* or *nocheztli*,[22] although other interpreta-

[4] Motolinía, 1941: p. 197; 1951: p. 258. See also López de Gómara (1552), 1954: **2**: p. 143 (in works on cochineal, the date is often incorrectly given as 1525); A Herrera (*ca.* 1600), 1934–1956: **5**: p. 492. Sahagún ([*ca.* 1570] 1963: pp. 123–124) described thirteen kinds of nopal; similarly Hernández (1571–1576), 1615: p. 30; 1651: pp. 76–78.

[5] Linné, 1948: p. 111.

[6] Griffiths and Hare, 1907: pp. 22–26; Diguet, 1928: pp. 136–139.

[7] Sartorius, 1961: p. 23. *Cf.* Oviedo y Valdés (1535), 1959*b*: **1**: p. 267 (*vino tinto*); Robelo, 1904: p. 616 (*nochote, nochocle*); Griffiths and Hare, 1907; p. 27; Diguet, 1928: pp. 139–142.

[8] Hernández (1571–1576), 1959: **1**: p. 312.

[9] Motolinía (1541), 1941: p. 197; López de Gómara (1552), 1954: **2**: p. 143.

[10] Harcourt, 1939: p. 60; Girault, 1966: p. 167 (*huarako, kichka*); Oblitas Poblete, 1969: p. 38 (*achuma*). *Cf.* Pena, 1901: p. 166 ("las hojas en apósito se aplican á las partes inflamadas"). From the Andes, Debelmas (1975: p. 107) reports the use of the joints, roots, trunk, and fruit of *Opuntia floccosa.*

[11] La Cruz (1552), 1940: p. 301; Schendel, 1968: p. 68. Also Berlin, Breedlove and Raven, 1974: p. 450 (Maya). *Cf.* Monardes (1577), 1925: **2**: p. 17 (*cardones*).

[12] Oviedo y Valdés (1526), 1959*a*; p. 99. *Cf.* Debelmas, 1975: p. 107; Ford, 1976: p. 297.

[13] Wight, 1850: **2**: p. 49, Chopra, Nayar and Chopra, 1956: p. 181 (India); F. P. Smith, 1871: p. 291, Wilson, 1913: **1**: p. 155 (China); Irvine, 1961: pp. 90–91 (Ghana); Watt and Breyer-Brandwijk, 1932: pp. 123–124 (southern Africa). Watt and Breyer-Brandwijk also state that Cactaceae are an ingredient in some native beers. Gerth Van Wijk (1911: p. 919) refers to the preparation of a gum from the branches of *O. tuna* and *O. vulgaris.*

[14] Perhaps properly described as a group of species. Brand (1966: p. 80) suggests that *O. hernandezii, O. tuna, O. megacantha,* and *O. decumana* may be placed within it.

[15] Earlier (1753) assigned to the genus *Cactus*, then (1768) *Opuntia*, and finally (1849) *Nopalea.* Humboldt (1811, **3**: p. 71) and Hooker (1827: pp. 2741–2742) pointed out that this was probably not the plant on which fine cochineal was raised (*cf.*, for example, Bryant, 1783: pp. 181–182). The error has, however, been repeated into modern times. Dunlop (1847; p. 124) discusses the kinds of cactus used in Guatemala.

[16] La Cruz (1552), 1940: p. 301 and pl. 90 (*tlatoctli*, planted); Hernández (1571–1576), 1959: **1**: p. 312.

[17] Motolinía (1541), 1941: p. 197; 1951: p. 258; Laet, 1633: p. 229 ("mirabilis arbor, quem Mexicani, Nochtli, insulares Tunam appellant [Europaei ficum Indica]").

[18] Clark (ed.), 1938: **1**: p. 19 n. 6; López de Gómara (1552), 1954: **2**: pp. 142–143; 1964: p. 157; Hernández (1571–1576), 1959: **1**: p. 311; Gage (1648), 1928: p. 71; Orozco y Berra, 1880: **1**: p. 267.

[19] La Cruz (1552), 1939: p. 38. Thus *nocheznupalli*, cochineal nopal (Sahagún [*ca.* 1570], 1963: p. 239). See also Hernández (1571–1576), 1615: pp. 31–32; 1651: pp. 76–79.

[20] Clark (ed.), 1938: **2**: pp. 17, 38, 40; Jiménez Moreno (ed.), 1940: p. 59. *Cf.* Peñafiel, 1885: **1**: p. 149; **2**: pl. xix. "Nopal" (*nopalli*) is also incorporated in a number of place names (Robelo, Olaguibel and Peñafiel, 1966: p. 127).

[21] Diguet, 1906: pp. 21–22; 1909: p. 76; 1928: p. 481; Reyes [1593], 1750: p. 158 (*atoco*). *Cf.* Alvarado, 1593, under *grana color afinada.*

[22] López de Gómara (1552), 1954: **2**: pp. 142–143; 1964: p. 157; Muñoz Carmargo (*ca.* 1576), 1948: pp. 234–235; *Anon.* [Manuscrit Ramírez, 1586–1589], 1903: pp. 26–27; Dorantes de Carranza (1604), 1902: p. 8; San Antón Muñón (*ca.* 1607), 1965: p. 55; Gage (1648), 1928: p. 71.

tions have been proposed.[23] The city was represented by an eagle perched on a nopal,[24] a subject drawn from Aztec legend and today the emblem of the Mexican nation. The nopal, together with maize, maguey (*Agave americana*), and the tule reed (*Scirpus* sp., rootstock and fiber plant) have been described as the "cuatro fuentes de bienestar" of Mexican rural life.[25]

The best-known varieties of *Opuntia Ficus-indica* were the nopal de San Gabriel (? *O. hernandezii*) and the more highly improved nopal de Castilla. In the province of Oaxaca, Mexico, the former was used in the regular production of cochineal, the latter for breeding purposes. Our knowledge of the raising of cochineal on the cultivated nopal during the colonial period rests heavily upon the accounts of two contemporary observers. The first, in time and in importance, is Gonzalo Gómez de Cervantes, whose *Vida económica y social de Nueva España* (*ca.* 1599 [1944]) includes several sections on *la grana cochinilla*. The other is José Antonio de Alzate y Ramírez who published a *Memoria en que trata del insecto grana ò cochinilla* in 1794–1795, the very year in which cochineal, albeit a wild Brazilian species, was first purposefully and successfully introduced to the Old World. In addition, a *memorial* of 1620 (addressed to the Viceroy of New Spain) and two *informes* (1776) by, respectively, Fr. Joaquín Vasco and Fr. Vicente Magán, priests in charge of townships in Oaxaca, supply valuable observations.

The nopal, while tolerating soils notably deficient in nutrients, responds well to improved conditions. Cultivated varieties benefited from the presence of wood ash on newly prepared plots (*rosadas*) and from the application of houseyard refuse.[26] Gómez advocated planting "en tierra muy labrada y beneficiada y que esté muy cultivada y tan limpia"[27] (fig. 2). Subsequent writers also stressed the importance of frequent cultivation and the removal of all other plants (*arboles, legumbres, semillas, flores*), both as a means of conserving moisture and fertility, and to reduce the risk of fungoid infections to which the nopal was prone. Irrigation was practiced during particularly dry spells (unlike some other cacti, the roots of the opuntia lie near the surface of the ground).

Nopals were normally propagated from the joints, cut cleanly from the old plant. To prevent rotting, the scar was first allowed to heal for fifteen to twenty days.[28] The *pencas* were then planted a half-meter deep, in rows two to three meters apart. Slopes shel-

FIG. 2. Cultivating a *nopalera* using a *coa*. Gonzalo Gómez de Cervantes (1599), 1944.

tered from cold winds and less liable to frost than open valley floors were preferred.[29] Good air drainage was also desirable during warm, humid weather, for under such conditions the insects were particularly susceptible to wasting diseases known in southern Mexico as *choreo* and *chamusco*.

A high proportion of cochineal came from small Indian holdings, but from the eighteenth century there were also large plantations (*haciendas*) of 50,000 nopals or more.[30] These were sometimes arranged in blocks of about twenty-five meters square. Surrounding mud walls or live hedges gave protection from wind and dust, and helped to exclude chickens and turkeys which devoured the *grana*.[31] Frost and driving rain also resulted in heavy loss of insects. Fires were sometimes lit when frost was expected,[32] and canopies of wood and straw (*tapextles*) might be erected to shield the plants from heavy downpours. Rafael Landívar (1781) advised that

> when a rainstorm is sweeping across the wide fields, or hail threatens to strike the colony, protect them with coverings as do the Indians, who throw wide mats over them every year. These clever people, having erected on this side and that timbers which exceed the height of the spreading nopals, fasten to these a large awning made of thick matting which they draw back and forth by means of a rope.[33]

---

[23] López de Gómara (1552), 1954: **2**: pp. 143–144; 1964: p. 158.

[24] Durán (1581), 1967: **2**: p. 63 and pl. 6.

[25] Robelo, 1905: p. 274.

[26] Alzate (1777–1794), 1852: p. 83 ("Suelen tambien abonarlo en Oajaca, ò plantar en tierras vìrgenes, despues de derribar y quemar el bosque . . .").

[27] Gómez de Cervantes, 1944: pp. 166, 171.

[28] Magán (1776), 1963: p. 76.

[29] Landívar (1781), 1948: p. 196; Alzate (1777–1794), 1852: p. 83; R. Blanchard, 1883: p. 80.

[30] Humboldt, 1811: **3**: p. 73. Stephens (1841: p. 294) remarked upon a plantation at Amatitlán, Guatemala, with 400,000 nopals.

[31] Gómez de Cervantes (1599), 1944: p. 169; Thiéry de Menonville, 1787: p. 302; Stephens, 1841: p. 294; Figuier, 1868: p. 135. *Grana* (grain) was earlier used to describe the related dye-insect kermes (for example, Pegolotti [*ca.* 1310–1340]; 1936: pp. 123 ff.; Pérez y López, 1797: **21**: p. 213 [1511]).

[32] Landívar (1781), 1948: p. 196; A. Alcedo (1786–1789), 1967: **4**: pp. 295–296.

[33] Landívar (1781), 1948: p. 196. See also Humboldt, 1811:

The extended arms of older plants were often supported by poles.

Nopals were ready for seeding with cochineal after eighteen months to three years, depending on the locality. From the fourth or fifth year, biennial pruning served to remove hard or decayed *pencas,* kept the height of the plant to a convenient two meters or so, and promoted new growth. The insects thrived best on fresh, tender leaves.[34] It was also considered beneficial to remove flowers and fruit in order to conserve the strength of the plant and to prevent the breeding of harmful insects.[35] The nopal, "tan doncella y delicada," was unfortunately subject to many pests and diseases. Gómez described various insects and worms that attacked the trunk, joints, and roots of the plant. Great care had to be exercised to keep the nopals clean, especially between harvests. The length of time that elapsed before a plantation had to be re-stocked depended on the standard of management, on the natural advantages of the region, and on the type of cactus grown. Ten to fifteen years might be soon enough in favorable circumstances. But during this time, especially on large holdings, plants were normally rested after every second or third year. In the words of Antonio de Alzate: "El consechero de cochinilla no se contenta con una sola nopalera; es necessario tener dos ó tres para pasar la cria de una á otra, y dejarlas descansar alternativamente, porque el nopal ha alimentado una generación queda ecschausto, ó como dicen los mexicanos, *quemado.*" [36] This form of rotation increased the cochineal harvest over the lifetime of the plantation. The methods employed in the *nopaleras* varied somewhat from region to region, but in general improved under the stimulus of high demand during the greater part of the colonial period. They were later applied, with modifications, in different parts of the Old World.

INSECT AND CULTURE

INSECT

Cochineal insects belong to the family Coccidae and to the genus *Dactylopius* (Costa). The latter gradually superseded *Coccus* (L.) or *Pseudococcus* (Westwood) as the generic description after the publication of M. E. Fernald's classification of the family in 1903.[37] Domesticated insects are usually assigned to a single species, *D. coccus,* the wild or partly domesticated insects to several species, although the differences and the respective distributions, particularly within the New World, have not been satisfactorily determined.[38] E. E. Green could find in *D. tomentosus, D. confusus,* and *D. indicus* "no sharp characters by which they may be defined." He was inclined to regard "these wild forms (including *D. argentinus* and two "sub-species") as comparatively recent derivations of *tomentosus*" (the smallest of the species). *D. coccus,* on the other hand, "has characters so distinct as to make it probable that its segregation dates from a much earlier period." [39]

From the time of our earliest knowledge of the New World, two main kinds of cochineal insect have been recognized: a domestic form, source of *grana fina* (*nocheztli*), and a wild form, from which *grana silvestre* (*xalnocheztli, ixquimiliuhqui*) was obtained. Antonio de Herrera (*ca.* 1600) referred to several varieties of *silvestre* (*salvagina, loca, montesina*).[40] All the wild species yield the characteristic dye, cochinealin or carminic acid (first isolated in 1818), but there are apparently considerable differences between them in the amount of pigment and, according to work on the South American species, less important variations in spectrographic range.

Dactylopius coccus is double the size of other species and takes about twice as long to complete its life cycle. Father Bernabé Cobo (1653) compared the full-grown insect (4 to 6 mm in length) to a chickpea or kidney bean (*garbanzo o frísol*).[41] The domestic form is also distinguishable by its coating of fine, waxy powder, whence the term "mealy bug" or "mealy cochineal"; the wild species, on the other hand, have a prominent cottonlike tomentum (thus "cotton cochineal") which may partly explain their much greater climatic tolerance. Humboldt found "wild cochineal . . . . in abundance in the most opposite climates" and in situations up to 2,900 meters.[42] This robustness was probably a matter of common knowledge, and can be inferred from references to the presence of *grana silvestre.* The distribution extended to areas that were notably colder rather than wetter than those in which the domesticated species thrived. An annual rainfall of less than 1,000

3: p. 73; Sartorius (1850), 1961: p. 172. Diguet (1928: pp. 514–515) describes two kinds of shelter, one permanent, the other temporary.

[34] Vasco ([1776], 1963: p. 47) observed that the cochineal of the Sierra de Oaxaca was "más gruesa" than in the Valle because the nopal of the former area had "más jugo nutricio."

[35] Gómez de Cervantes (*ca.* 1599), 1944: p. 111 ("el tunal que criare tunas, criará poco cochinilla"); Sartorius (1850), 1961: p. 172.

[36] Alzate (1777–1794), 1852: p. 85.

[37] Linnaeus established the genus *Coccus* (Lat. *coccum*). The Spanish word *cochinilla* was possibly derived from *coccum* (Hernández [1571–1567], 1959: **1**: p. 315) or from *coccineus,* scarlet-coloured (André, 1949: pp. 116–117). However, *cochinilla* can also mean a woodlouse (which the New World insect rather resembled) or, more generally, any insect parasite, and the name may have been transferred. See Richter, 1701: pp. 2–8 (*De nomine cochinillae*).

[38] A. Alcedo ([1786–1789], 1967: **4**: p. 295) described *cochinilla* as "Género de insecto del orden cleoptera; . . . . hay 49 especies, que se distinguen por el número y color de las manchas de sus alas, y por las plantas en que viven."

[39] Green, 1912: pp. 79–80.

[40] A. Herrera, 1934–1956: **9**: p. 155; also León Pinelo (1650), 1943: p. 247.

[41] Cobo, 1956: **1**: p. 202; *Memorial* (1620), 1931: p. 498 (*garbanzo pequeño*).

[42] Humboldt, 1811: **3**: p. 76; similarly, Caldas y Tenorio (1810), 1912*a*: p. 432. See also Ruusscher, 1730: p. 265.

mm and a dry season of three to four months were close to the minimum requirements.

The question whether *D. coccus* derives from a superior wild form which has now disappeared, or is a product of the improvement of one of the known wild species cannot be satisfactorily resolved. The notable differences between the present wild and cultivated forms at first suggest the former. Early commentators, however, generally assumed that the two were directly related, the differences being ascribed solely to cultivation.[43] Certainly, the domestic *Dactylopius* is a delicate and vulnerable insect, closely dependent on man. Improvement has come about through the selection and care of breeding populations, management of the host nopals, and partial protection from a large number of enemies, besides inclement weather. Gómez de Cervantes described in some detail the creatures that devoured or attacked cochineal larvae and the immobile insects (fig. 3). They included turkeys and chickens (*gallinas de la tierra, gallinas de Castilla*), birds generally, and various pests that called for constant surveillance—lizards (*lagartijas*), a kind of leech (*sanguijuela*), an insect "resembling a spider," [44] and several grubs or worms (*gusanos*).[45] Rats and mice, armadillos, and some snakes were similarly harmful.[46] In well-tended gardens, wild cochineal was removed,[47] primarily to conserve the nopals and also to guard against interbreeding. The wild population must always have been kept severely in check by natural enemies, and it is difficult to visualize an improved and less robust species surviving at all. *D. coccus* tends quickly to disappear when cultivation and protection cease.[48] The differences between *fina* and *silvestre* forms appear to be largely the result of processes of domestication operating over a considerable period of time.

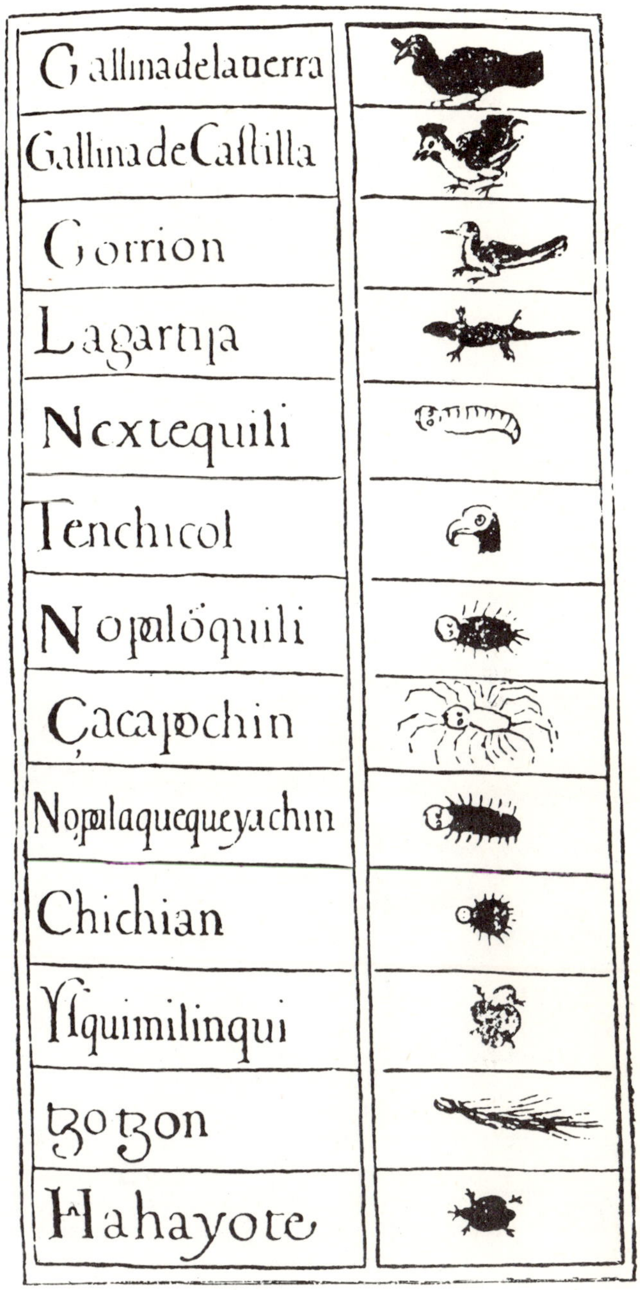

Fig. 3. Enemies of cochineal. Gonzalo Gómez de Cervantes (1599), 1944.

CULTURE

The cycle of events associated with the cultivation of cochineal varied somewhat in different parts of the New World,[49] chiefly according to the length and

[43] *Memorial* (1620), 1931: p. 50 ("la silvestre fue el mismo genero que el de la fina y en la cultivación está la diferencia"). Discussed by Humboldt (1811: **3**: pp. 65–66) and Dahlgren de Jordan (1963: p. 12).

[44] According to Diguet (1928: p. 520), spiders themselves were welcomed, chiefly because they fed on harmful insects, but not cochineal.

[45] Gómez de Cervantes (1599), 1944: pp. 172–174. He also gives the Náhuatl names. *Cf.* Vasco (1776), 1963: p. 53; Magán (1776), 1963: p. 81.

[46] Clavijero (*ca.* 1780), 1964: p. 233; Alzate (1777–1794), 1852: p. 84; Humboldt, 1811: **3**: p. 74. See also León Pinelo (1650), 1943: p. 248; Landívar (1781), 1948: p. 195; Dunlop, 1847: p. 130; Squier, 1858: p. 523; M. Herrera, 1919: p. 120.

[47] Vasco (1776), 1963: p. 46; Magán (1776), 1963; p. 76; Alzate (1777–1794), 1852: pp. 84–85; Humboldt, 1811: **3**: p. 75. *Cf.* Sahagún (*ca.* 1570), 1938: **3**: p. 288 (*grana falsa, ixquimiliuhqui*).

[48] Alzate ([1777], 1794: p. 228) refers to the testimony of the *alcade mayor* of Nexapa (southern Oaxaca), according to which *grana requisima* was to be found, seven leagues or so from Nexapa, without cultivation, either of nopals or insects. This has no clear explanation, and falls short of proof of the existence, then or earlier, of a superior wild species.

[49] Remarked by Ulloa (1748), 1760: **1**: pp. 341–347; but

FIG. 4. Harvesting cochineal, *ca.* 1700 (*Archivo General de Indias,* Sevilla). R. C. West and J. P. Augelli, 1966.

severity of the wet and dry seasons. Domesticated insects were ready for harvesting after 90 to 120 days. The same nopal might therefore be seeded twice or even thrice in the course of a year. On large nopalries, rows of plants were seeded (and harvested) at intervals of about ten days so as to make the best use of labor. Seeding (*asemillada*) was usually accomplished by attaching to the joints of the nopal "nests" or *cartuchos* each containing ten to twenty-five pregnant insects (*madres*).[50] The nests were made of a variety of materials, in some areas an epiphyte resembling moss, called "Spanish moss," *barba de español* or *paxtli* (*Tillandsia* sp.),[51] in others palm fiber (*zacate de palma*), maize sheaths (*tlasole*), or bark. Insects were also placed in hollow reeds or cane (notably in Guatemala)[52] or in bags (*bolsitas*) of muslin or open calico. The latter custom was adopted by the cochineal farmers of the western Mediterranean and of the Canary Islands.[53] Rather than several small colonies of *madres,* a larger nest with 100 or more insects was sometimes moved from one joint to the next.

Newly hatched insects (*semilla, peojillia*) leave the nest and gradually spread over the adjacent joint; then, after embedding their proboscis in the tissue of the plant, they become completely immobile. The mother insects perish in the nests; collected and dried, they were marketed as *grana de paxtle, zacatillo,* or *cascarilla.*[54] In Guatemala in the middle of the nineteenth century, insects reared around Amatitlán from October (near the end of the wet season) were mostly used to seed the nopalries of Antigua, about ten miles away, in January.[55] Amatitlán produced two crops of cochineal annually (the second made into dyestuff), Antigua only one.

Wild cochineal might be collected, from untended nopals, throughout the year. For the domesticated species there were harvest seasons, usually two or three (including the collection of *zacatillo*) per annum. The impregnated females double in size in the course of their short life. They were cropped before depositing eggs, leaving only sufficient *madres* to provide *semilla* for the second main harvest. The insects were carefully removed from the joints of nopal with a quill, a pointed stick (*cuchara*), or some kind of switch or brush (*zilhuastle*).[56] The harvester worked with an instrument in one hand and a shallow, rimmed receptacle (*chilcalpetl*) in the other (fig. 4). The receptacle was usually of wood, less commonly of plaited straw or earthenware (*vasijas vidriadas*).

A distinctive feature of cochineal culture was the rearing of insects under cover during "winter" (*invierno*). Joints of nopal well populated with *madres* were

most descriptions are based on only one area and give the impression that timing and procedures were the same everywhere.

[50] Female insects outnumber the males by about 200 to 1, and only the females were harvested.

[51] Jiménez de la Espada (ed.) (1580's), 1965: 2: p. 268 (*telaraña*); Ajofrín (1760's), 1964: 2: p. 121 (*pascle*); Raynal (1770), 1776: 2: p. 206; Alzate (1777–1794), 1852: p. 84 ("ciertas plantas parasíticas, llamadas *paxtles magueitos*"); Humboldt, 1811: 3: p. 73. On *Tillandsia usneoides,* see La Cruz (1552), 1940: p. 298 and pl. 86; Robinson, 1947: pp. 47–49; Martínez, 1959: p. 484 ("*paxtli* o heno," literally "hay"). Clavijero ([*ca.* 1780], 1964: p. 240) refers to "nidos de heno," i.e. *paxtli.*

[52] Vasco (1776), 1963: p. 64 (*nidos de cañaveral*); *Reports,* 1839: p. 3; Stephens (1841), 1846: 1: p. 277; Dunlop, 1847: p. 129.

[53] Mackenzie, 1831: p. 212; O. M. Stone, 1889: p. 204; Morris, 1895: p. 97. Pitard and Proust (1909: p. 197) state that *Opuntia tuna* was introduced from America to provide needles (*épines*) to attach the *bolsitas.*

[54] Usually regarded as equal to or better than the main harvest. Vasco ([1776] 1963: p. 48) reported from Oaxaca that 1 lb of *grana pastle* was equivalent to 1½ lbs of standard cochineal. See also Dunlop, 1847: p. 127; Nobrega, 1848–1849: p. 345; Squier, 1858: p. 522; M. Herrera, 1919: p. 171.

[55] Dunlop, 1847: p. 127. Haefkens ([1827], 1969*b*: pp. 46–47), Dunn (1829: pp. 158–159) and Squier (1858: 521–523) also include useful eyewitness accounts of the industry in Guatemala.

[56] Gómez de Cervantes (1599), 1944: p. 168; Alzate (1777–1794), 1831: p. 281; Diguet, 1928: p. 517.

cut and stored at the onset of heavy rain. First, they were placed in closed baskets or under the roofs of houses, and the newly hatched *semilla* remained dormant for about three weeks.[57] The joints were then exposed to light and air in shelters fitted with racks or compartments. On large plantations, long, narrow buildings (*almacenes*) were sometimes constructed for this purpose.[58] Under cover, the insects reached the stage of propagation in about four months when, the wet season largely over, they could be placed in nests and taken out to the live nopals to begin a new cycle of production.[59] If, for some reason, insects were not reserved from the last harvest of the year, or a new nopalry was being established, joints complete with *semilla* could be purchased. Humboldt referred to this trade in the market of Oaxaca, center of an important cochineal-producing area throughout the colonial period.[60] He also described from the same province a curious and perhaps unique variation on the routine of cultivation:

The Indians who cultivate the cochineal and who go by the name of *nopaleros,* especially those who live round the town of Oaxaca, follow a very ancient and very extraordinary practice, that of making the cochineal travel. In that part of the torrid zone, it rains in the plains and vallies from May to October, while in the chain of neighbouring mountains called Sierra de Istepeje, the rains are only frequent from December to April. In place of preserving the insect in the rainy season in the interior of their huts, the Indians place the mother-cochineals, covered with palm-leaves by beds in baskets made of very flexible claspers. These baskets (*canastos*) are carried by the Indians on their backs as quickly as possible to the mountains of Istepeje, above the village of Santa Catalina, at nine leagues distance from Oaxaca. The mother-cochineals produce their young by the way. On opening the *canastos* they are found full of young *coccus,* which are distributed on the nopals of the sierra. They remain there till the month of October when the rains cease in the lower regions. The Indians then return to the mountains in quest of the cochineal for the purpose of replacing it in the nopaleries of Oaxaca. The Mexican in this way withdraws the insects from the pernicious effects of the humidity in the same manner as the Spaniard travels with his merinos from the cold.[61]

PREPARATION OF COCHINEAL

Cochineal was an unusual product so far as the New World was concerned. R. M. Gilmore, writing of South America, observed that "aboriginal man in Neotropica did not process any animal product except possibly red dye from cochineal insects; i.e., he did not tan hides into leather by chemical means, prepare perfumes from animal musk, make glue from animal gelatin, or soap from fats." [62] This is also true of Middle America, except that a wax was prepared from another species of insect, *Llaveia axin.*

Cochineal varied in quality with the prevailing physical conditions, the standard of cultivation and, to an important extent, the method of killing the insect. The last evoked considerable discussion throughout the colonial period. Gómez de Cervantes was sufficiently interested to experiment with five methods, reporting on the loss of weight (usually about one-third) and the time and labor involved.[63] The simplest, and possibly the oldest, method was to spread the insects on mats placed in the sun.[64] The results were most satisfactory if the "baking" was periodically discontinued or the mats were partly shaded, but this meant that the operation might last for anything up to two weeks. Nevertheless, sun-drying was usually considered best [65] and produced a fine silver cochineal (*plateada, blanca*); in 1573 the Spanish authorities attempted—without success—to prohibit other methods.[66] The use of heated rooms reduced the time involved to several hours, followed by four or five days in the sun. These rooms were the ancient *temazcalli,* designed for hot vapor baths.[67] Dry ovens were also used. From Guatemala, R. G. Dunlop in 1847 reported "stoves . . . each capable of containing 100 to 200 baskets." [68] By this time, too, oven-drying had been introduced to the Canary Islands [69] and to Java.[70] Given the correct temperature, the product was an ash-gray cochineal, known in Mexico as *jaspeada* (marbled). When the insects were killed and dried on hot plates (the *comales* employed in making *tortillas*),[71] a black cochineal (*negra*) was produced. Yet another dry method was to mix the insects with hot, powdered ashes, referred

[57] Landívar (1781), 1948: p. 196; Clavijero, 1787: 2: p. 381; A. Alcedo (1786–1789), 1967: 4: p. 296; Humboldt, 1811: 3: p. 73.

[58] Dunlop, 1847: p. 125 (Amatitlán); Maudslay and Maudslay, 1899: p. 27.

[59] O'Neale (1945: p. 30) has an account of keeping insects (and making small amounts of cochineal) through the wet season in Jocotenango, Guatemala.

[60] Humboldt, 1811: 3: p. 73.

[61] Humboldt, 1811: 3: pp. 77–78. Later descriptions (for example, Tylor, 1861: p. 24) are based on Humboldt. There may be an allusion to the custom in Magán (1776: p. 76) and in *Instrucción declaratoria sobre repartiento de la grana por los alcades mayores* (Dahlgren de Jordan (ed.), 1963; p. 97).

[62] Gilmore. 1950: p. 346.

[63] Gómez de Cervantes (1599), 1944: pp. 174–176. See also *Voto Consultivo del Real Acuerdo,* in Dahlgren de Jordan (ed.), 1963: pp. 88–93.

[64] Gómez refers to the practice (apparently usual in the Mixteca) of first shaking the insects in a small jug (*cántaro*).

[65] Ulloa (1748), 1760: 1: p. 345; Clavijero, 1787: 2: p. 381; 1964: p. 233; A. Alcedo (1786–1789), 1967: 4: p. 296; Biart, 1885: p. 194.

[66] Lee, 1948: p. 470.

[67] Ruusscher, 1730: p. 266; Clavijero (*ca.* 1780), 1964; p. 233; Alzate (1777–1794), 1831 (illustration); Humboldt, 1811: 3: p. 78.

[68] Dunlop, 1847: p. 134. *Cf.* O'Neale, 1945: p. 30.

[69] O. M. Stone, 1889: p. 133. Nobrega (1848–1849: p. 347) noted that the heat required was similar to that for baking bread.

[70] Itier, 1848: 2: p. 340.

[71] Ruusscher, 1730: p. 266; Alzate (1777–1794), 1852: p. 86; Clavijero (*ca.* 1780), 1964: p. 233; R. Blanchard, 1883: p. 76. The method was often preferred when only small quantities of *grana* were being prepared.

to by Cervantes de Salazar (1554) and Antonio de Herrera (*ca.* 1600).[72]

A common alternative procedure involved the use of boiling water or steam. The insects were placed either in bags and then submerged, or in a pot with a small amount of boiling water, or in cylinder-like baskets (*tenates, tompeates*) made to fit over the neck of a steaming cauldron.[73] Some of the insects usually burst, giving the cochineal, when later dried, a brown, dull red or rosy black color. Alzate described the product as *denegrida,*[74] which was generally considered less desirable than *grana* prepared by one of the dry methods. Whether anything was gained by the use of cold water or cold water and vinegar [75] is not recorded, but the custom seems to have been unusual. It probably originated in the Old World where kermes insects were often placed in a solution of vinegar and water. Another unusual method was reported (1693) from Jamaica where cochineal insects, without being removed from the joints of nopal, were suffocated by smoke.[76] The creatures could then be shaken onto cloths placed around the base of the nopal. In the Canary Islands, at the beginning of the twentieth century, insects collected in muslin bags were sometimes dispatched with sulphur fumes.[77]

Grana could easily be adulterated [78] and the practice of adding foreign matter, or of mixing *fina* and *silvestre,*[79] was of constant concern to the authorities. As early as 1548, the *cabildo* of Tlaxcala, Mexico, instructed Indian officials to supervise the trade in cochineal.[80] In 1572, Viceroy Martín Enríquez appointed a *juez de grana* in the city of Puebla; later, there were *jueces* in Oaxaca, Tlaxcala, and Cholula. The purpose was to try to ensure that the dyestuff that left New Spain was genuine and properly described. Bernardino de Sahagún (*ca.* 1570) referred to cochineal mixed with chalk or powdered clay (*greda*) or flour; this was *grana baja* or *mezclada,* known in Náhuatl as *tlapalnextli* "from *tlapalli* ("color") and *nextli* ("ashes"), because this same cochineal is a preparation, just a mixture. Some has chalk, some has flour, some is a mixture [of both]." [81] Adulterants also included pitch, sand, and various other minerals—chosen according to the color of the *grana*—as well as fragments of vegetation accidentally added in harvesting the insects. The best cochineal was sifted after drying to remove such fragments and any small or broken insects (*granilla, mostacilla*). The latter could be sold separately but fetched only a fraction of the price of the standard product.[82] Finally, cochineal might be falsely increased in weight, and its durability impaired, by inadequate drying or by dampening before packing. The opportunities for malpractice were indeed numerous, but what took place in New Spain itself was apparently unimportant in comparison with what often happened after the dyestuff was landed in Europe.[83]

In the *Matrícula de Tributos* (1511–1512+), a copy of which was included with the *Codex Mendoza* (1540's), cochineal from Oaxaca and the Mixteca is represented by bags (*talegas*) with red spots [84] (fig. 5). Under the Spaniards, too, *grana* for export was packed in bags (*zurrones*) for loading on mules. Before shipment it was transferred to boxes (*cajones*) or bales covered with hides.[85]

For local use, the Indians made cochineal into cakes or tablets (*panes, panecitos, pastillas*), known as *nocheztlaxcalli.*[86] The *Codex Florentino* shows such cakes being made and displayed for sale (fig. 6C). The method of preparing the paste was apparently something of a trade secret [87] and may also have varied

[72] Cervantes de Salazar, 1953: p. 61: A. Herrera, 1934–1956: **9**: p. 155; also *Memorial* (1620), 1931: p. 498.

[73] Alzate (1777–1794), 1831 (illustration).

[74] Alzate (1777–1794), 1852: pp. 85–86; Figuier, 1868: p. 138 (*ronagrida*); Schützenberger, 1867: **2**: p. 337; Diguet, 1928: p. 524 (*renegrida*); Born, 1938: p. 216 (*regenerida*). See also Vetancurt, 1698: p. 189; Ruusscher, 1730: p. 266; Ulloa (1748), 1760: **1**: p. 345; Landívar (1781), 1948: p. 197; A. Alcedo (1786–1789), 1967: **4**: p. 296.

[75] A. Herrera (*ca.* 1600), 1934–1956: **9**: p. 155; Alzate (1777–1794), 1852: p. 85.

[76] *Anon.,* 1693: pp. 502–503. The English naturalist John Ellis (1758), acting on behalf of Linnaeus, hoped to obtain live insects from Jamaica, but was informed that "[the inhabitants of the island] do not cultivate that plant (opuntia) for this purpose"; he therefore made enquiries, ultimately with successful results, in Carolina (Ellis, 1763: p. 662; J. E. Smith, 1821: **1**: pp. 106–107, 118, 131; *cf.* Adrosko, 1968: p. 24). *Nopalea cochenillifer* (L.) is listed in the *Hortus Eastensis* ([1789] 1806) of the Botanical Garden, Liguanea, Jamaica (Powell, 1972: p. 64).

[77] Sage, 1907: p. 140.

[78] Gómez de Cervantes (1599), 1944: pp. 176–180; *Memorial* (1620), 1931: pp. 500, 502–504.

[79] A. Herrera (*ca.* 1600). 1934–1956, **9**: p. 155; León Pinello (1650), 1943: p. 247.

[80] Gibson, 1952: p. 150. Lee (1948: p. 457) refers to an investigation into adulteration in 1554.

[81] Sahagún, 1963: pp. 239–240; *cf. ibid.,* 1961: p. 77, and 1938: **3**: pp. 287–288.

[82] A customs account drawn up at Vera Cruz in October, 1790 (Dahlgren de Jordan (ed.), 1963; Ap. I—1) distinguishes *grana fina, granilla, polvo de grana,* and *grana silvestre.*

[83] Humboldt, 1811: **3**: p. 79. Born (1938: pp. 224–245) describes the kinds of falsification practiced in the importing centers, including the manufacture of artificial cochineal. Re-exported cochineal was particularly suspect.

[84] Clark (ed.), 1938: **1**: pp. 78–79: **3**: folios 43–45; Lorenzana, 1770: fig. 23; Peñafiel, 1890: 2, 2: nos. 248, 249.

[85] McCulloch, 1832: p. 283; Dunlop, 1847: p. 134; Sartorius (1850), 1961: p. 172.

[86] Sahagún, 1963: p. 239; also *ibid.,* 1961: p. 77, and 1938: **3**: p. 287 ("véndenla en los *tiánquez* [markets] hecha en panes").

[87] "En la provincia de Tlascala . . . . se hace el carmín . . . . y no quieren los naturales descubrir el secreto de cómo se cuaja." (A. Herrera [*ca.* 1600], 1934–1956: **9**: p. 155). In 1552, cochineal *en panes* was reported as tribute from Autlán in Jalisco, "la única parte donde parece ser conservado el arte de hacer panes de grana," according to Dahlgren de Jordan (1963: p. 14). Villaseñor y Sanchez (1748: **2**: p. 208) included

FIG. 5. Representing 40 bags (*talegas*) of cochineal. *Codex Mendoza and Matrícula de Tributos* (1511–12 + [copied 1540's]), 1938.

between different regions. According to Francisco Hernández (*ca.* 1575), milled cochineal was added to a liquid in which alum and *tézhoatl* leaves had been boiled, and the sediment that collected was molded into *pastillas.*[88] This was the refining process (*purificada*) mentioned by Sahagún.[89] The cakes were, however, inconvenient to transport over long distances, and the compound itself was unsuitable for the European market. Humboldt observed that "separated *grana*" was preferred so that "the Indians may not introduce extraneous matter in those agglutinated masses called *bodoques* (pellets, balls)."[90] A kind of cake-cochineal

FIG. 6. Upper—*nopalli* and *nocheztli;* Center—*tlaquauac tlapalli:* cleaning, selecting and painting with cochineal; lower—*nocheztlaxcalli:* cakes (*panes*) of cochineal being prepared and displayed for sale. Bernardino de Sahagún *Codex Florentino* (*ca.* 1570), 1963.

was also prepared in Peru where it was known as *macnu,* and there are parallels in the preparation of *bixa* in Middle and South America[91] and of lac dye in South and Southeast Asia.[92]

---

grana in torta among the products of Autlán. *Cf.* Miranda, 1952: p. 262.

[88] Hernández, 1959: **1**: p. 315 ("Se obtiene sumamente exquisito si se muele el cocimiento del árbol llamado *tézhoatl,* del que en su lugar hablaremos, agregándole alumbre y recogiendo el sedimento, que se guarda en forma de pastillas"). For *tézhoatl, tézhuatl,* or *tesguate* (*Miconia laevigata*), see *ibid.* **2**: pp. 118–119. The addition of alum, apart from its usefulness in dyeing, may have prevented or checked decomposition. For a discussion of the problem of preserving cake-cochineal, see *Reports,* 1839: p. 109. Pure and untreated cochineal kept almost indefinitely (Savary des Bruslons [1723], 1751: p. 519; Hellot [1789], 1901: p. 102).

[89] Sahagún, 1938: **3**: p. 287; 1963; p. 242.

[90] Humboldt, 1811: **3**: p. 79. Ray (1688: **2**: p. 1465) and Sloane (1725: p. 153) also refer to *grana en pan.* The Spaniards had elaborate regulations concerning the preparation and use of dyestuffs (Pérez y López (ed.), 1797: **21**: pp. 203–233, especially 222).

[91] Donkin, 1974: p. 40 (*panecillos, pelotas, bollas*).

[92] K'ang-hsi (*ca.* 1700), 1779: p. 478 (*pains*).

FIG. 7. The painter. *Codex Mendoza and Matrícula de Tributos* (1511–12 + [copied 1540's]), 1938.

USE OF COCHINEAL

Pigments with a cochineal base were available in Mexico in the sixteenth century [93] and probably long before. *Nocheztli* and *achiotl* were apparently ingredients in the common red-brown paint used in codical art.[94] The *Codex Florentino* (*ca.* 1570) describes the preparation and blending of colors [95] and shows an illuminator at work with several cakes of *nocheztli* by his side (fig. 6B). Cochineal was employed as a cosmetic by Indian women [96] and also in decorating various kinds of craftwork.[97] Bernal Díaz, companion of Cortés, recalled seeing in the market of Tlaltelolco rows of "gaily painted jars (*jarros muy pintados*) made of wood." [98] Until late in the eighteenth century, the potters of Olinilá (Guerrero) used *grana* mixed with other substances.[99] The Zoque of Chiapas painted their houses with *grana*.[100] Modern reports refer to the use of cochineal paint on baskets (the Lenca, southern Honduras) and on poles carried by dancers (Izalaco, Guatemala).[101]

The earliest Spaniards commented enthusiastically upon the range of colors of Mexican fabrics. There was "spun cotton, in all colors, so that it seems quite like one of the silk markets of Granada, although it is on a larger scale." [102] Although only a few fragments of pre-Conquest date have survived,[103] it is apparent that cochineal was used in the dyeing of cloth, at least from the basin of Mexico southward to the isthmus of Tehuantepec.[104] Representations of red cloaks in some Mixtec codices probably closely matched the material itself, the coloring agents being identical or very similar. Several early colonial sources refer to *grana para tintoreros*.

Cotton, the only superior fiber available in ancient Middle America, could not be satisfactorily dyed with cochineal alone. A mordant was required to make the color fast, and for this alum was employed in Middle and South America, in the latter for both cotton and woolen fabrics. Sahagún observed that alum (*tlalxocotl*) was well known in Mexico.[105] Father Cobo (1653) mentioned "four or five kinds of alum" used in dyeing,[106] and Hernández described how alum was an ingredient in *pastillas* of cochineal. Later, the dye was occasionally mordanted with the juice of limes

---

[93] Motolinía (*ca.* 1550), 1970: p. 86; Roys, 1931: p. 51 (Yucatán).

[94] Clavijero (*ca.* 1780), 1964: p. 249; Boban, 1891: **2**: p. 211; Burland, 1955 (*Selden Roll*): p. 11, 1966 (*Codex Laud*): p. 11; Adelhofer, 1963 (*Codex Vindobonensis Mexicanus*): p. 25; Dahlgren de Jordan, 1963: p. 14; Donkin, 1974: p. 34. *Cf.* Nowotny and Strebinger, 1958 (*Codex Becker I*): p. 226. Ochre, cinnabar, and colcothar were also used. Vaillant (1940: p. 49) states that cochineal pigments were occasionally employed by fresco artists (Yucatán, Oaxaca). A. G. Miller (1973: p. 178) identified only haematite at Teotihuacán. See also E. Edwards, 1966: p. 9.

[95] Sahagún, 1938: **3**: pp. 287–290; 1963: pp. 239–245.

[96] Sahagún, 1938: **3**: p. 47; 1961: p. 55; A. Herrera (*ca.* 1600), 1934–1956: **9**: p. 155 ("el carmín para los pintores y para las mujeres"); Lee, 1948; p. 453; Rojas Gonzaléz, 1949: p. 81 (Nicaragua). Red paint applied to the face or body on ceremonial occasions was very common in the New World and many substances were employed, including ochre and achiote (*Bixa orellana*); see, for example, Thompson (1923), 1973: pp. 77, 79 (Maya); Dahlgren de Jordan, 1954: p. 126 (Mixtec).

[97] Brasseur de Bourbourg, 1858: **3**: p. 31.

[98] Díaz del Castillo, 1956: p. 217: 1963; p. 195.

[99] Alejo de Meave (1791), 1831; p. 218. We know that the township was producing cochineal in 1610 (Mota y Escobar, 1945: p. 264). In Europe, cochineal-lacquer became a speciality of Florence and later of Paris and Vienna (A. Anderson, 1764: **1**: p. 306; Merrifield, 1849: **1**: pp. clxxvi–clxxvii, 645; Born, 1938: p. 225).

[100] Dahlgren de Jordan, 1963: p. 14.

[101] D. Stone, 1948: p. 211; Osborne (1933), 1965: p. 169.

[102] Cortés (1520), 1908: **1**: p. 258. See also *Anon.* (1520's), 1917: pp. 29–30; Tamara, 1556: p. 274; Zorita (*ca.* 1575), 1964: pp. 66–67.

[103] Cordry and Cordry, 1968: pp. 6, 178, 252; Johnson, 1954: pp. 137–147.

[104] *Memorial* (1620), 1931: p. 497; Keleman, 1944: **1**: p. 215; Cook de Leonard, 1971: p. 220.

[105] Sahagún, 1938: **3**: p. 289; 1963: p. 243 and fig. 817 (alum being dug with a *coa*). See also La Cruz (1552), 1940: p. 34.

[106] Cobo, 1956: **2**: p. 113 ("De piedra alumbre se hallan en estas Indias cuatro o cinco especies. En la Nueva España lo hacen muy blanco, lúcido y transparente, del cual usan los tintoreros en sus tintes . . . .")

or oranges.[107] It is also reported that *grana* was mixed with other dyestuffs, either as a reinforcing agent (with shellfish purple) [108] or to give different shades of red.[109] Wool and silk were dyed with cochineal in Mexico from the middle of the sixteenth century.[110] In the state of Oaxaca, surviving items of traditional dress, such as *huipiles* and belts, sometimes incorporate cochineal-dyed silks; [111] and the dyeing of wool with cochineal apparently has not entirely ceased.[112] On the other hand, only one reference to the staining of leather has been found (p. 29 *infra*). The practice seems to have originated and to have been largely confined to the Old World, notably southern Spain and Northwest Africa.

B. Dahlgren de Jordan claimed that the "medicinal properties" of cochineal were well known to the ancient Mexicans.[113] Certainly cochineal was an item in the *materia medica* of Middle America in the sixteenth century. Francisco Hernández, physician to Philip II and our chief authority, observed that, ground and mixed with vinegar, it was applied to wounds, used to clean the teeth, and to relieve ailments of the head, heart, and stomach.[114] That such "cures" were largely based on Indian recipes seems very likely. R. L. Roys has published two Mayan medical texts that refer to cochineal (*mukay*).[115] From at least the early seventeenth century, cochineal specifics were also prescribed in Europe.[116] They were included among the uses of cochineal in dictionaries of commerce in the eighteenth century,[117] and J. Pereira's *Elements of Materia Medica* (1840) mentions "several [unproven] medical properties." [118] Today the product is employed in medicine simply as a coloring agent. The use of cochineal to color foodstuffs in modern times probably has an ancient counterpart. In the region of Ometepec (Guerrero) cochineal is added to *tamales*.[119] Perhaps the custom developed in areas remote from supplies of *achiote*, the more common alimentary dye.

2. MIDDLE AMERICA

PRE-HISPANIC

Whether the use of cochineal in Middle America dates from Post Classic or Classic times, or even earlier, can at present be only a matter of speculation. One scholar has suggested that the dyestuff was known in Classic Teotihuacán through trade connections with the valley of Tehuacán.[1] Humboldt, for some unknown reason, associated it with the Toltecs.[2] By far the most important evidence takes the form of a tribute list, the *Matrícula de Tributos* (1511–1512+).[3] From this we discover that *nocheztli* was being levied by the Triple Alliance (Mexico, Texcoco, and Tlacopan) in conquered territories to the south of the basin of Mexico. The respective provinces, comprising twenty-five principal townships (Appendix A), were Tlaxiaco and Coixlahuaca in the Mixteca (*Mistecapan*) and Cuilapan in the adjacent highlands of central Oaxaca.[4] Together they were responsible for supplying sixty-five *talegas* or *cargas* (9,750 lbs) [5] annually. Other sources indicate that cochineal was demanded elsewhere in Oaxaca and produced in parts of Tlaxcala (map. 2). In northeastern Oaxaca, the province of Tuxtepec had to deliver to Texcoco "400 pieces of cloth *de grana*." [6]

Cochineal may have been collected or even cultivated very widely in western and south-central Mexico in pre-Columbian times. The dyestuff was used locally and also traded. Merchants carried it into southern-

[107] Hardy, 1829: p. 455 (the Río Babispe—Río Guachinero area of northern Mexico); Osborne, 1935: p. 54, 1965: p. 41 (Momostenango, Guatemala); Start, 1963; p. 18. In Europe, from the time of Cornelius Drebbel (1572–1634), chloride of tin was used to give a vivid scarlet. Cochineal dyes a different red according to the mordant used, with alum crimson.

[108] Covarrubias, 1946: p. 246.

[109] Vázquez de Espinosa (*ca.* 1620), 1942: p. 655; Osborne, 1965: p. 41.

[110] Motolinía (*ca.* 1550), 1970: pp. 6–7; A. Alcedo (1786–1789), 1967: 1: p. 295.

[111] Cordry and Cordry, 1968: pp. 58, 77, 101, 103, 140, 168. In India, lac continued to be used chiefly to dye silk after the introduction of aniline compounds (O'Conor, 1876: p. 57).

[112] Cordry and Cordry, 1968: p. 214 (Otomí).

[113] Dahlgren de Jordan, 1961: p. 391.

[114] Hernández (1571–1576), 1959: 1: p. 315. Simpson (1938) has published an exceptionally early (1536) American pharmacopoeia; the three New World ingredients do not, however, include cochineal.

[115] Roys, 1931: p. 337. *Coccus axin* was used in the preparation of a yellow ointment.

[116] J. Gerard [T. Johnson], 1633; pp. 1512–1513; Pomet, 1709: p. 13; Geoffroy, 1736: p. 370; *Pharmacopoeia*, 1771: p. 52; 1788: p. 14; 1825: p. 237; absent, however, from Pemel (1652–1653) and Lemery (1698).

[117] Savary des Bruslons (1723), 1751: 1: p. 518; A. Anderson, 1764: 1: p. 306.

[118] Pereira, 1839–1840: 2: p. 1378. Hooper (1910: p. 510) refers to *Coccus cacti* (Linn.) in the *materia medica* of India; but he may have confused cochineal with kermes or with some other insect of western Asia (the vernacular names *kirmaj*, *kirmdāna* [Hind.] and *karmiz* [Pers.] are given).

[119] Dahlgren de Jordan, 1961: p. 391.

[1] Cook de Leonard, 1971: p. 219.

[2] Humboldt, 1811: 3: p. 69.

[3] The history of the *Matrícula* is discussed by Barlow (1949) who also identifies and locates the constituent townships. The most complete version of the *Matrícula* is a copy made on the authority of the first viceroy, Don Antonio de Mendoza in the 1540's (Clark [ed., 1938], *Codex Mendoza*, Part 2). See also Miranda, 1952: pp. 23–28; Molíns Fábrega, 1954: pp. 303–335; Scholes and Adams, 1957; Borah and Cook, 1963: pp. 28–33.

[4] The armies of the Triple Alliance invaded the Mixtec-Zapotec country in the middle of the fifteenth century.

[5] Dahlgren de Jordan (1961: p. 391) supplies this calculation on the assumption that the *talega* corresponded to the colonial *zurrón* (of *grana*) of 150 lbs.

[6] Alva Ixtlilxochitl (*ca.* 1605), 1892: p. 198; Brasseur de Bourbourg, 1857–1859: 3: p. 267. Barlow (1949: p. 96) translates "cuatro-cientos paños de grana" as "400 measures of cochineal." Other provinces rendered cloth patterned in red (*ibid.*: pp. 53, 56, 59, 62).

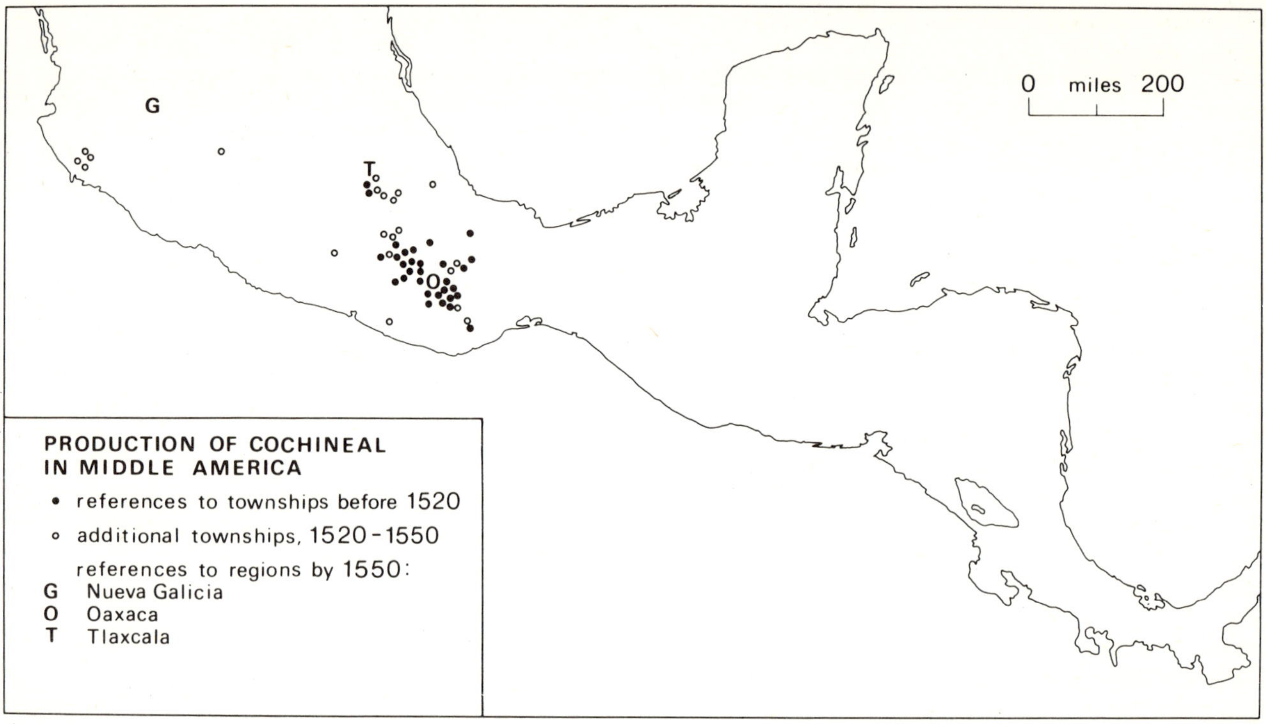

MAP 2. Cochineal in Middle America (—1520 to 1550).

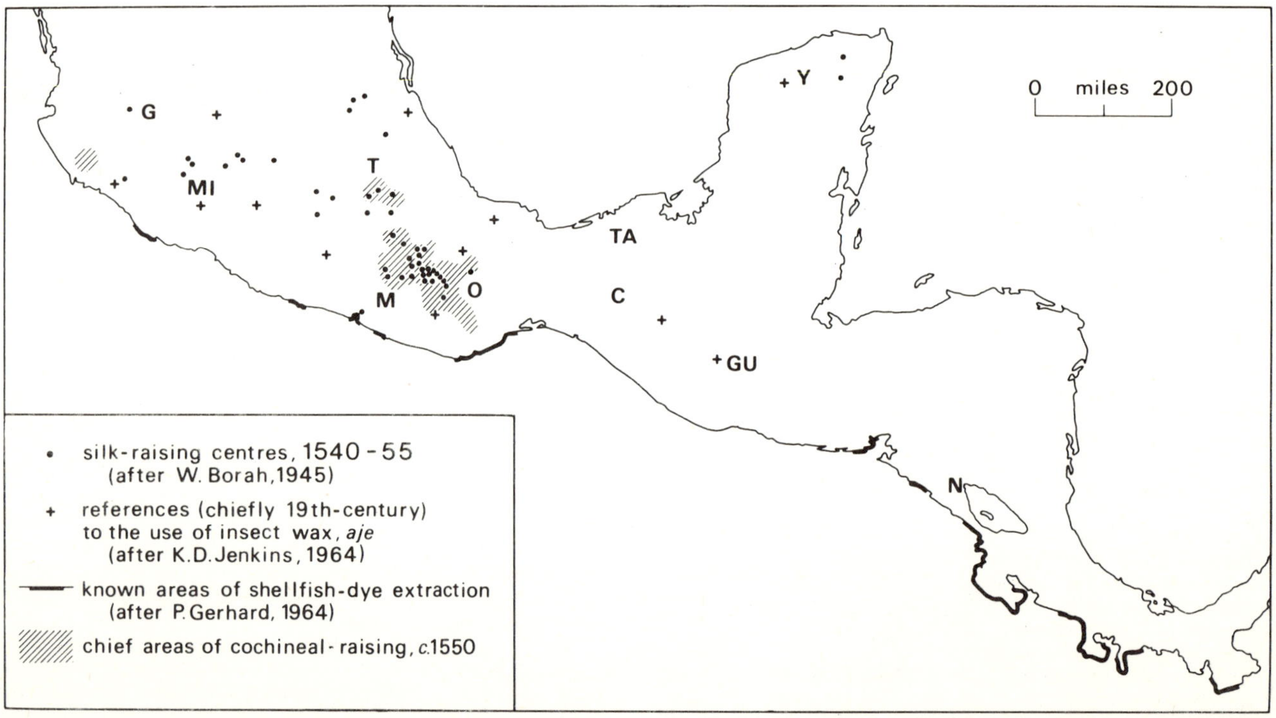

MAP 3. Cochineal- and silk-raising (*ca.* 1550), shellfish-dye extraction, and insect-wax production in Middle America.

most Mexico and Guatemala.[7] Nevertheless the tribute list almost certainly identifies the chief areas of specialized production at the beginning of the sixteenth century. The Mixteca (forty-five *talegas*) stands out prominently. This and later evidence have fostered the belief that the region was the primary center of cochineal culture.[8]

The peoples of Middle America exploited other animal species for industrial purposes—shellfish for purple-red dye and insects for wax (map 3). Rather similar skills and levels of environmental perception were involved in these activities, which may be historically and functionally related. In particular, two forms of insect culture, found in the same general area and with similar or complementary products, are unlikely to have developed independently. The wax insect *Llaveia* [*Coccus*] *axin* was almost certainly known in pre-Columbian times.[9] The archaeological evidence is not conclusive,[10] but *axin* (Náhuatl) and *ni-in* or *niih* (Maya) are reported from the sixteenth century, notably by Diego de Landa (*ca.* 1566) and Bernardino de Sahagún (*ca.* 1570).[11] In his *Relación de las cosas de Yucatán,* Landa writes of "a red worm from which a good yellow ointment is made . . . by only crushing them or kneading them together; and it serves as oil for painting vessels and makes colors fast."[12] Sahagún refers to medical prescriptions and to cosmetic preparations.[13] The "wax" (*aje*) is not a secretion (like Chinese *pe-la* and lac resin) but body fat of the insect itself. *Llaveia axin* was probably carried as a domesticate from the *tierra caliente* of southern Mexico and Guatemala into areas farther south.[14] The Guaymí of Panama, and others who cultivate *axin,* seed the host trees by attaching colonies of eggs,[15] somewhat after the methods employed in rearing cochineal. A connection between the insect cultures of Middle America, if and when established, may be paralleled in Southeast Asia where the exploitation of lac and of *pe-la* appear to have a common ancestry.

[7] Brasseur de Bourbourg, 1857–1859: **3**: p. 240; Roys, 1943: p. 107. *Cf.* Emmart, 1940: p. 32.

[8] Clavijero (*ca.* 1780), 1964: p. 42; Brasseur de Bourbourg, 1857–1859: **3**: p. 31.

[9] Diguet, 1906: p. 37; Rojas González, 1940: p. 34; West, 1948: p. 68; Cook de Leonard, 1971: p. 220 (? pre-Classic).

[10] Jenkins, 1964: p. 631.

[11] Also Palacio (1576), 1866: p. 20; and, among seventeenth-century authorities, Nieremberg, 1635: p. 289; Vetancurt (1698), 1971: p. 52.

[12] Landa, 1941: p. 193. *Cf.* La Cruz (1552), 1940: p. 32; Roys, 1931: p. 337 and medical texts nos. 253, 337.

[13] Sahagún, 1961: p. 90. Some modern authorities mention a red dye prepared from *Coccus axin* (Boban, 1891: **2**: p. 211; Edmonson [annotating the *Popol Vuh*], 1971: p. 79). The contemporary Tarahumara, who acquire *axin* (Tara., *alí*) through trade, use it "only as a relish with chile" (Bennett and Zingg, 1935: p. 130).

[14] Gordon, 1957: p. 47.

[15] *Ibid.,* p. 38; Jenkins, 1964: p. 627.

COLONIAL

1520–*ca.* 1550

A very early, if somewhat puzzling, reference to cochineal is contained in the instructions issued, in Spain, to Diego de Sámano on his appointment as a royal official in the province of Vitoria Garayana (Pánuco). The unpublished document, dated 20 December, 1523, refers to reports of *grana* in the province—well outside the known area of pre-Columbian production—and to the arrival of consignments in Spain.[16] Officers of the crown were ordered to inform themselves of the amount of dye produced and of the possibilities for the future. Antonio de Herrera (*ca.* 1600) mentions a similar report of 1523 concerning "New Spain," followed by instructions to Cortés himself.[17] Cochineal was readily obtainable in Tenochtitlán. The *conquistador* Bernal Díaz, writing about 1568, remembered seeing "mucha grana" on sale in the great market of Tlaltelolco,[18] and it is possible that samples of cochineal were among the first Mexican products to be shipped to Europe. Alternatively, the knowledge that the peoples of the mainland possessed a superior red dyestuff (probably confused with kermes) may have reached Spain with the arrival of native fabrics in 1519 or 1520.[19]

The *memorial* on cochineal incorporated in a *real cédula* of 1620 opens with the observation that at the time of the Spanish conquest the Indians of New Spain produced "poca cantidad la grana cochinilla fina."[20] This can best be understood in the light of the rapid expansion of cochineal farming in the second half of the sixteenth century. On the other hand, Antonio de Herrera was clearly in error in stating that, while the Indians knew cochineal in 1519, they paid little attention to it and were taught the arts of cultivation by the Spaniards.[21] The Dominicans, whose province in

[16] Chipman, 1967: p. 100.

[17] A. Herrera, 1934–1956: **7**: p. 25 ("Y habiendo tenido el Rey noticia que en Nueva-España nacía grana en abundancia, y que traída a Castilla podía redundar en mucho provecho para las Rentas Reales, mandó al Gobernador que lo mirase e hiciese coger, y avisase luego, si esto era verdad, y que le parecía que para beneficiarla se podía hacer.").

[18] Díaz del Castillo, 1963: p. 195.

[19] Cortés (1520), 1908: **1**: pp. 170–171. The gifts from Moctezuma II to Cortés included "five or six thousand pieces of rich cotton stuffs, woven and embroidered in divers ways" (*ibid.,* p. 234). Similarly, Zorita (*ca.* 1575), 1864: pp. 66–67 (". . . mucha ropa . . . . que considerando ser toda de algodon y sin seda, en todo el mundo no se podia hacer ni tejer cosa tal, ni tantas, ni tan diversos y naturales colores; . . . ropas de hombre y mujer muy maravillosas . . . que hechos de seda no se podian comparar. Y habia otros paños como tapiceria . . .").

[20] *Memorial,* 1931: p. 497. *Cf.* Gómez de Cervantes (1599), 1944: p. 164 ("Esta grana cochinilla se criaba y cogía en el tiempo de la infidelidad de los indios en poca cantidad porque no tenian salida de ella, y sólo les servía para dar tinte a algunas mantas de poca consideración . . .").

[21] A. Herrera (*ca.* 1600), 1934–1956: **9**: p. 154 (". . . pues aunque los indios la tenían, no hacían caso de ella, que los

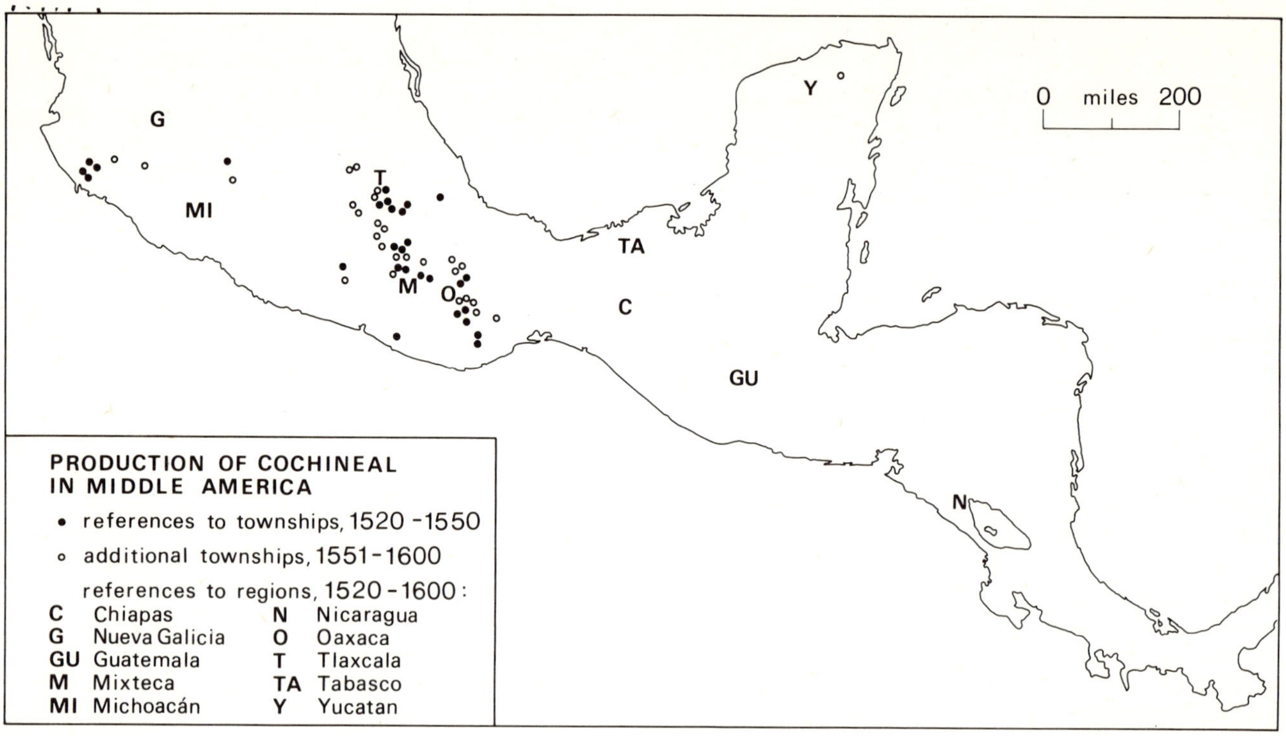

Map 4. Cochineal in Middle America (1520 to 1600).

the sixteenth century broadly coincided with southeastern Mexico, are known to have stimulated production from the 1540's. This is probably what Francisco de Burgoa, a member of the order, meant when he wrote that the friars taught (*enseñaron*) the people of Tlapalcatepeque in southern Oaxaca to plant *nopaleras de grana.*[22] At Coixlahuaca Fr. Francisco Marín (*ca.* 1550) was apparently instrumental in extending the area devoted to the cultivation of both cochineal and silk.[23] Perhaps the wealth to be derived from *grana fina* was an important consideration in founding missions in the region of Oaxaca, including the Mixteca.

The Spanish government gave little direct encouragement to cochineal production in the first half of the sixteenth century, and what action was taken appears to have been confined to the province of Tlaxcala. The president of the second *audiencia,* Don Sebastían Ramírez de Fuenleal (1531–1535), ordered that the Indians of the area around the town should apply themselves to the rearing of cochineal.[24] By 1548 nominees of the *cabildo* were supervising the trade in *grana* which was valued at about 200,000 pesos annually. Official interest in Tlaxcala, rather than the Mixteca, may partly be explained by the comparative ease with which commodities could be moved to Vera Cruz, the principal port of shipment throughout the colonial period.

Grana was incorporated into the royal tribute system from at least the 1530's. Private *encomenderos* too exacted cochineal at an early date.[25] Demand was rising and, although no figures are available, we can probably assume that production kept pace with this and that an increasing number of townships cultivated the insect. The total amount of cochineal marketed through official and private channels from the mid-1530's was probably substantially in excess of that circulating *ca.* 1520. Information about the townships harvesting *grana* between 1520 and 1550 comes chiefly from the *Suma de Visitas* of 1548–1550. The distribution (maps 2 and 4), while undoubtedly incomplete, tends to confirm more general statements and to suggest some expansion in the areas of specialized production. Within the dominant, central zone, Tlaxcala was

castellanos les han enseñado . . ."). Likewise, Antonio de Solis Rivadeneyra, 1684: pp. 181–182 (". . . la cochinilla, cuyo uso no conocian, hasta que le aprendieron de los Españoles . . ."). Denied by Clavijero (*ca.* 1780), 1964: p. 42.

[22] Burgoa (1674), 1934: **2**: p. 289. *Cf.* Bustamante, 1821: p. 18.

[23] Dávila Padilla, 1625: p. 242 ("Y para que huuiese de que sacarlas, les mandò plantar nopales de grana, y morales para coger seda . . .").

[24] Torquemada (1615), 1969: **1**: p. 608 ("Ordenò, que en Tlaxcalla, y todos sus alrededores se pusiese cuidado en la criança de la Grana, ò Cochinilla"). Similarly, A. Herrera (*ca.* 1600), 1934–1956: **9**: pp. 247–248 ("Que en Tlascala se pusiese cuidado en las crianzas de la cochinilla, con lo cual, y con estar reservada de tributo la provincia de Tlascala y por el mercado franco . . .").

[25] Lee, 1948: p. 454 (places in Oaxaca and Vera Cruz); Scholes, 1936: p. 201 (Huajuapan in the Mixteca). Concerning the supply of cochineal under *repartimiento* arrangements in the seventeenth and eighteenth centuries, see Scholes and Roys, 1948: p. 305 (Yucatán); Hamnet, 1971*b*: p. 76 (Oaxaca); W. B. Taylor, 1972: p. 147 (Oaxaca).

apparently more important than in 1520. There are also isolated reports from Michoacán (Huaniqueo) and Vera Cruz (Cocula) and a notable cluster around Autlán (later known as Autlán de la Grana) in Jalisco. Perhaps the Franciscans were responsible for initiating production in western Mexico,[26] but it is equally probable that, like the Dominicans in the south and east, they merely encouraged an interest in cochineal that antedated the conquest. *Grana color afinado* was known in Tarascan as *charapeti yuscata.*[27] Herrera, using a report on New Galicia (Zacatecas) of 1531, states that although *grana* was abundant in the province, no attention was then paid to cultivation.[28]

1550–*ca.* 1600

Cochineal farming at this time was entirely in the hands of the native population. The unit of production was the *huerta* rather than the *hacienda* and individual producers, particularly *caciques* with larger holdings (*cacicazgos*), grew rich on the sale of *grana.*[29] So much so that in 1551 and 1553 the Indian *cabildo* of Tlaxcala, alarmed at the effect of unaccustomed affluence on traditional forms of authority, attempted, albeit unsuccessfully, to prohibit further planting of nopals.[30] The central government, on the other hand, was more interested in expanding production, and the striking growth of exports in the final decades of the century owed much to the efforts of two viceroys, Martín Enríquez de Almansa (1568–1580) and Pedro Moya de Contreras (1583–1585).[31] *Alcades mayores* (Spanish officials in charge of districts) were ordered to extend cultivation; and to combat fraud and regulate trade generally a *juez de grana* was appointed (1572) in the city of Puebla. Provincial governors who actively promoted the industry included Gonzalo Gómez de Cervantes in Tlaxcala and, somewhat later (1612), Antonio de Figueroa in Yucatán and Campeche.[32] Gómez, in his important account of cochineal (1599), described the product as "almost equal [in value] to silver" in the trade of New Spain.[33] In this he greatly exaggerated, but it is nevertheless true that the dyestuff ranked second to precious metals at the end of the sixteenth century. Proposals for a royal monopoly in cochineal were circulating before 1600.

In Oaxaca and the Mixteca cochineal was comparatively neglected during the middle decades of the sixteenth century in favor of silk. Between the two insect cultures there were obvious parallels, and the conflict of interest can be inferred from the respective distributions (map 3). Silk-rearing was introduced in the 1530's and developed strongly in the 40's and 50's, particularly in the Mixteca.[34] Again the Dominicans promoted cultivation, with considerable effect until the close of the century when supplies from China, shipped through Manila, began to dominate the market. Silk-rearing in Oaxaca declined in importance after about 1580, but never entirely ceased. Nor was this the only part of Mexico where both cochineal and silk were raised. They were found together in Puebla, to some extent in Michoacán, and probably also in Chiapas.[35]

Sources for the second half of the sixteenth century include the extant *Relaciones Geográficas* (1577–1580), reports based on a questionnaire which, under item 25, called for information on the raising of cochineal. The Tlaxcala-Puebla region occupied a leading position over the period as a whole; and towards the close of the century, interest in *grana* in Oaxaca and the Mixteca began to revive. The English traveler John Chilton referred particularly to three townships in the vicinity of Puebla (Tepeaca, Huejotzingo, and Tecamachalco) whence "they bring most of their *cochinilla* into Spain" (1568). And again of Oaxaca (1570): "All the Indians of this province pay their tribute in mantles of cotton wooll, and *cochinilla,* whereof there groweth abundance."[36] *Grana* was one of the products that drew merchants from all over central Mexico to the celebrated market at Coixlahuaca in the Mixteca Alta.[37] The main cochineal zone consisted of a belt of country between central Tlaxcala and southern Oaxaca (maps 4 and 5) and at the turn of the sixteenth century no one part dominated the whole. There were also more remote producing areas, including some unknown or only inferred in 1550. Communities on the margins of the valley of Mexico collected or raised cochineal, but expansion was chiefly to the south and east of the Isthmus of Tehuantepec—into Chiapas, Tabasco, Yucatán, Guatemala, and Nicaragua—where the industry

26 Brand, 1966: p. 82.

27 Gilberti (1559), 1901.

28 A. Herrera (*ca.* 1600), 1934–1956: **9**: p. 234 (". . . mucha grana cochinilla, aunque no se beneficia . . ."), p. 238.

29 *Ibid.* **5**: p. 145 (Guazocingo [Huejotzingo]). *Cf.* Burgoa (1674), 1934: **2**: p. 289 (Tlapalcatepeque); Dahlgren de Jordan, 1963: p. 18 (Miahuatlán, Ocelotepec, 1580).

30 Gibson, 1952: p. 149.

31 Gómez de Cervantes (1599), 1944: p. 164; *Memorial* (1620), 1931: pp. 497–498. Viceroys Luis de Velasco II (1590–1595) and Gaspar de Zuñiga y Acevedo (1595–1603) also exerted some influence; see Bancroft, 1883: p. 621 (1597).

32 Gómez de Cervantes (1599), 1944: p. 164; *Memorial* (1620), 1931: p. 501.

33 Gómez de Cervantes (1599), 1944: p. 163. *Cf.* Torquemada (1615), 1969: **3**: p. 384 ("la mejor mercadería de Indias").

34 Borah, 1943: p. 15. *Cf.* Motolinía (*ca.* 1550), 1970: pp. 6–7; Tomson (1555–1558), 1904: p. 358 (". . . there is a place called Misteca . . . which doth yield great store of very good silke and cochinilla . . ."). Bartolomé de Zárate in his *Relación descriptiva del Valle de Oaxaca* ([1544], 1915: pp. 253, 256) mentions *seda* but not *grana.*

35 Motolinía (1541), 1951: pp. 326–327; Gage (1648), 1928: p. 167; Bazant, 1964: pp. 478–482. Markham (1892: p. 493) refers to similar developments in the Arequipa region of Peru.

36 Chilton, 1904: pp. 363–364. Concerning Tlaxcala and the neighboring provinces of Calpan and Huejotzingo, see also Scholes and Adams (ed.) [1562], 1958: p. 96.

37 Durán (1581), 1967: **2**: p. 185.

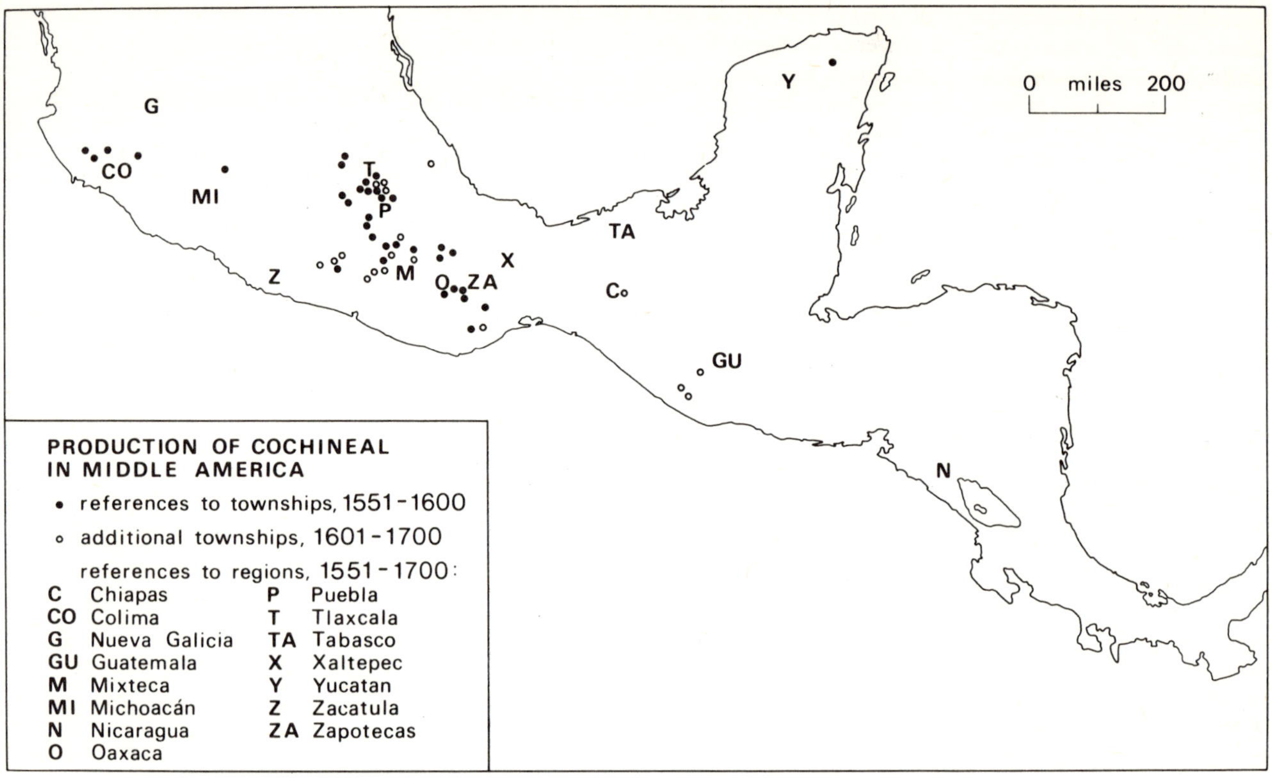

MAP 5. Cochineal in Middle America (1551 to 1700).

probably had ancient antecedents.[38] According to a report of 1579, *grana* was available in the Guatemalan province of Zapotitlán-Suchitepéquez.[39] Small quantities of cochineal from Guatemala and Nicaragua reached Seville in 1581 and 1590 respectively.[40] In these borderlands the product was *grana silvestre* rather than *grana fina.* Attempts to introduce the latter, or to promote cultivation of the indigenous insect, apparently had very limited success.[41]

1600–*ca.* 1800

Central and Southern Mexico

The country around Tlaxcala and Puebla produced substantial quantities of cochineal until at least the middle of the seventeenth century. Juan de Torquemada (1615) observed that crowds gathered in the Saturday market of Tlaxcala for the disposal of "muchas arrobas de grana." [42] Of the same city, Vázquez de Espinosa wrote: "It takes in quantities of fine cochineal, as do other cities and villages in its jurisdiction; and if the Indians paid tithes on it, as the bishop proposes and has taken legal steps to authorize, the diocese will have an annual income equal to that of the archdiocese of Toledo." [43] The Sierra de Tlaxcala was "the best stretch of land to be found in New Spain; they collect great quantities of fine cochineal here"; and "in many of the villages [of the Sierra between Puebla and the city of Mexico] they raise and prepare the finest cochineal." [44] Cholula, too, drew considerable wealth from the sale of *grana.*[45] León Pinelo (1650), after naming the chief centers of production in New Spain, including Oaxaca and the Mixteca, added that Tlaxcala occupied first place ("pero mas y mejor en el distrito de Tlaxcala").[46] This does not exactly accord with other contemporary statements, but there was probably little or no decline in output until the second half of the seventeenth century. The reasons for the decay of the industry in this area after *ca.*

[38] Landa (*ca.* 1566), 1966: p. 132; Molina Solis, 1896: p. 853; Roys, 1943: pp. 51, 110.

[39] Estrada and Niebla (1579), 1955: p. 77. President Villalobos had earlier encouraged the planting of nopals in this region. There is, however, no reference to cochineal in Juan de Pineda, *Descripción de la Provincia de Guatemala* (1549), 1925.

[40] Chaunu and Chaunu, 1955–1959: **6**: p. 981.

[41] Dahlgren de Jordan, 1961: p. 392; Macleod, 1973: p. 172.

[42] Torquemada (1615), 1969: **1**: p. 276.

[43] Vázquez de Espinosa (*ca.* 1628), 1942: p. 135 (based on observations between 1609 and 1612). *Cf.* Laet, 1633: p. 229. Crops of New World origin and grown on Indian holdings were usually exempt from tithe. The question whether cochineal raised on Spanish holdings (or holdings leased from Spaniards) was subject was in dispute until the end of the colonial period (Borah, 1941: pp. 386–409, 1949: pp. 498–517; Hamnet, 1971*b*: pp. 60–61).

[44] Vázquez de Espinosa, 1942: pp. 137, 166.

[45] *Ibid.*, p. 384: Mota y Escobar (1610), 1945: p. 248; Torquemada (1615), 1969: **1**: p. 282.

[46] León Pinelo, 1943: p. 248.

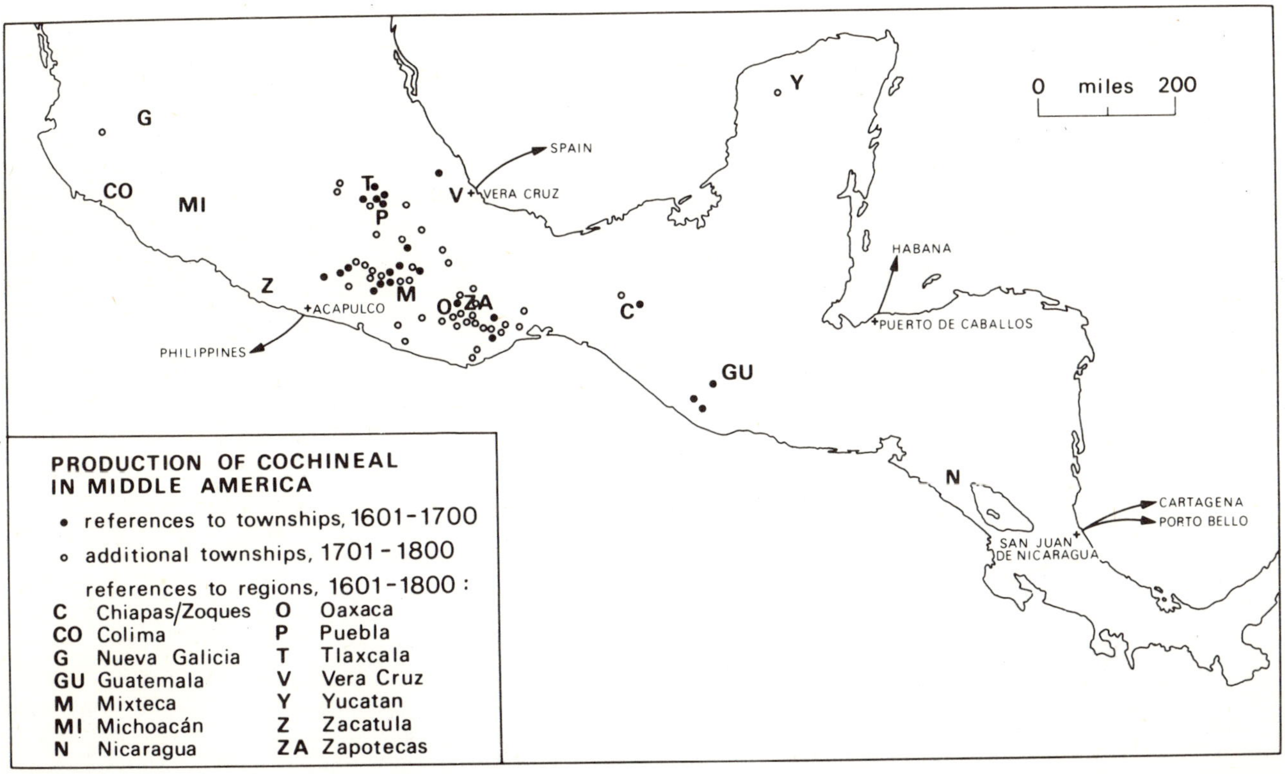

MAP 6. Cochineal in Middle America (1601 to 1800).

1650 are not entirely clear. Excessive use of the nopalries over the period of a century may have been one reason; but the chief was probably the small profit that accrued to the majority of the Indian cultivators while, at the same time, merchants and intermediaries, including the *alcades mayores,* prospered from the trade in *grana.*[47] The rising costs of administration were passed on to the producers and some communities were driven to destroy their plantations in order to escape exploitation.[48] By the late eighteenth century the industry was insignificant. *"Grana,"* wrote Clavijero (*ca.* 1780), "used to be cultivated in Tlaxcala, Huejotzingo and neighboring places." Humboldt described the rearing of cochineal as "a branch of industry which [the intendancy of Oaxaca] formerly shared with the provinces of Puebla and New Galicia."[49] Contemporary references to the cochineal of Tlaxcala in, for example, Antonio de Alcedo's *Diccionario Geográfico Historico de las Indias Occidentales ó América* (1786–1789), imply no more than the survival of cultivation.

Modest quantities of cochineal were harvested in provinces to the west and northwest of the basin of Tlaxcala-Puebla. In Michoacán, according to Torquemada, "criase grana, de la buena" (presumably *fina*).[50] Townships in the vicinity of the city of Mexico produced both *fina* (for export) and *silvestre* at the close of the eighteenth century.[51] Elsewhere the product seems to have been mainly, if not exclusively, *silvestre.* The most frequently mentioned region was New Galicia (or Guadalajara). Alzate (1794) put the output at only one-fortieth of that of Oaxaca, or less than 1,000 *arrobas* (25,000 lbs).[52] Vázquez (*ca.* 1612) remarked that the Indians of the coastal provinces of Colima and Zacatula "gather quantities of corn, cotton, and cochineal."[53] The same could probably have been said of much of western and northwestern Mexico; remote from the ports through which the dyestuff was exported to Europe, insects were collected chiefly for local use.

Between the early seventeenth and the late eighteenth centuries, the cultivation of *grana fina* became increasingly associated with southwestern Mexico, from the continental divide on the north and east to the Pacific coast on the south and west (maps 5 and 6). In almost all other parts of Middle America the industry appears to have steadily declined. Occasionally from 1600 and invariably after 1650 Oaxaca was described as the chief supplier of cochineal;[54] by the second half of the eight-

[47] Clavijero (*ca.* 1780), 1964: pp. 42–43.
[48] Dahlgren de Jordan, 1963: p. 17.
[49] Humboldt, 1811: 2: p. 241.
[50] Torquemada (1615), 1969: 1: p. 336.

[51] Dahlgren de Jordan, 1963: Ap. I (1); Gibson, 1964: pp. 354 (around Chalco), 569 n. 139 (Xochimilco). See also Gage (1648), 1928: p. 43.
[52] Alzate, 1852: p. 86; Humboldt, 1811: 3: p. 64 ("scarcely 800 arrobas . . . in a year"). *Cf.* Laet, 1633: p. 280 ("coccinallam minus laudatam" in New Galicia).
[53] Vázquez de Espinosa, 1942: p. 177.
[54] For example, *Memorial* (1620), 1931: p. 498 ("Las pro-

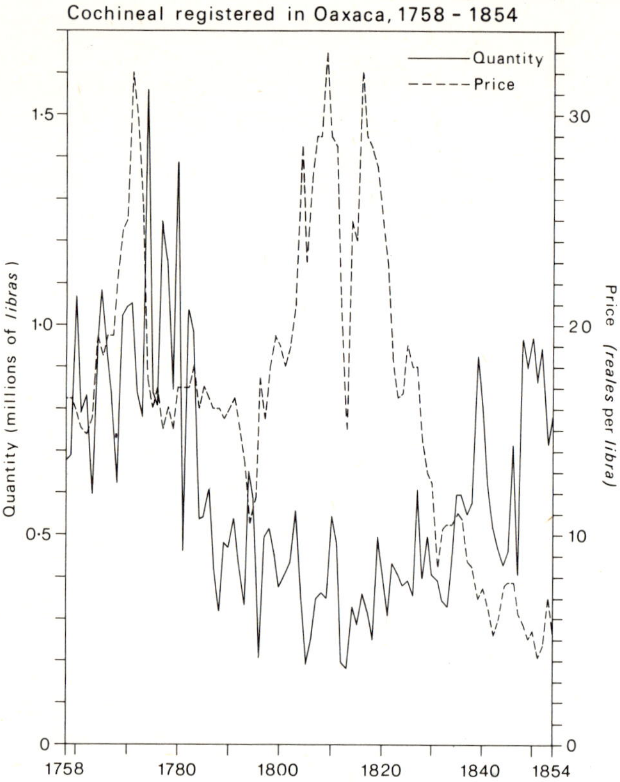

FIG. 8. Cochineal registered in Oaxaca (1758 to 1854).

eenth century it was the only region where significant quantities of *fina* were harvested.[55] Indeed the best cochineal was sometimes known in Europe as *mistec* or *mixteca*.[56] The notion that the name referred to Meteque in Honduras [57] can be traced perhaps to the shipment of *grana* (from Central America) through Puerto de Caballos. The Mixteca, more than other parts of southwestern Mexico, specialized in cochineal; a rugged and semi-arid province, large areas were more suitable for nopals than for food crops.

In Mexico the rearing of cochineal on large plantations commenced in the eighteenth century. This mode of production was, however, always exceptional and also appears to have been confined to the Southwest. "Around the town of Oaxaca and especially near Ocotlan," wrote Humboldt, "there are *haciendas* which contain 50 to 60,000 nopals . . . [but] the greatest part of the cochineal which is employed in commerce is . . . produced in small nopaleries belonging to Indians of extreme poverty." [58] *Hacendados* were discouraged by the uncertainties of production and the sharp variations in prices; by the large number of field laborers required, particularly at certain times of the year; and by the rather complex preparation of *grana fina* for the market. At the same time, larger holdings brought little saving in time and effort. The industry was peculiarly dependent on the skill and patience of individual workers, qualities generally encouraged by the prospect of personal gain.

For cochineal produced in New Spain there were three sources of demand. An unknown but doubtless relatively small quantity of *fina* and *silvestre* was used in native craftwork. Secondly, *grana*, presumably chiefly *fina*, was employed in Spanish-controlled woolen mills (*obrajes*). Manufacture commenced early in the 1540's [59] and regulations dating from the 1560's and 1570's restricted dyeing with cochineal to Spanish artisans and to cloth of high quality and full width. That no more than 200 *arrobas* of *grana* were used in New Spain in 1575, the view of Viceroy Enríquez,[60] must be regarded as a conservative estimate; and the amount surely increased until at least the middle of the eighteenth century. At no time, however, was the total large in comparison with the quantity exported, mainly to Spain, but also to the Philippines.

Shipments of *grana* varied considerably from year to year (p. 37 *infra*), as did output. The latter is shown, more or less directly, by the amounts registered in Oaxaca between 1758 and 1854 (fig. 8).[61] Such fluctuations can be accepted as characteristic of earlier times too. As we have seen, cochineal was highly vulnerable to adverse meteorological conditions: to sudden rainstorms, high winds, and unseasonable frosts. An entire crop might be lost in a matter of minutes. Precautions (not always taken) could only reduce the risk. Secondly, any notable increase in Indian mortality was especially significant for a labor-intensive industry. The effect on production and prices was first remarked in the *memorial* of 1620.[62] The influence of the major epidemics of the second half of the eighteenth century (1766, 1780, 1785–1787) can be detected in the cochineal registration figures from

vincias de la Mixteca y Oaxaca, que es a donde se coge la mejor . . ."); Laet, 1633: p. 261; Cobo (1653), 1956: **1**: p. 202 (". . . y donde hay más heredades della es en la diócesis de Guajaca . . .").

[55] Clavijero (*ca.* 1780), 1964: p. 42; A. Alcedo (1786–1789), 1967: **4**: p. 296; Alzate (1777–1794), 1831: p. 267 ("La grana fina solo se cultiva en el obispado de Oajaca . . ."), 1852: p. 83. Villaseñor y Sánchez (1746–1748: **2**: pp. 111–202) mentions *grana* among the products of many places in the bishopric of Oaxaca.

[56] Clavijero (*ca.* 1780), 1964: p. 43; Humboldt, 1811: **3**: p. 65 ("the mealy, fine or Mixtec cochineal"); Napier, 1853: p. 385.

[57] Sloane, 1707–1725: **2**: p. 208; Ferchault de Réaumur, 1734–1742: **4**: p. 90; Diderot and d'Alembert, 1751–1757: **3**: p. 560; Schützenberger, 1867: **2**: p. 335.

[58] Humboldt, 1811: **3**: p. 77; Clavijero (*ca.* 1780), 1964: p. 43. Concerning the condition of the Indians, see *Instrucción declaratorio sobre repartimiento de la grana por los alcades mayores* (Dahlgren de Jordan, 1963: pp. 97–100).

[59] Dusenberry, 1947: p. 223.

[60] Lee, 1948: p. 472.

[61] Tabulated in Dahlgren de Jordan, 1963: Ap. II: *Datos presentados por la oficina del registro y la administración principal de rentas de Oaxaca.* The same figures (up to 1819) are given in Bustamante, 1821: pp. 20–24.

[62] *Memorial,* 1931: p. 501.

Oaxaca. Thirdly, poor maize harvests upset the balance of regional and interregional trade whereby cochineal was regularly sold to buy corn. The neglect of food crops is mentioned as early as 1579,[63] and it was certainly a feature of parts of Oaxaca in the eighteenth century.[64] The relationship between the two economies, subsistence and commercial, was very unstable. Famines naturally resulted in higher maize prices which, in turn, led to increased planting and, for a time, a fall in the output of cochineal. Such cyclical trends varied in magnitude and territorial scope, sometimes affecting the greater part of southwestern Mexico. Finally, the market for *grana* was occasionally depressed by the disruption of overseas trade during periods of war between Spain and other European powers. Fr. Joaquín Vasco, parish priest of Santa María Ecatepec (southern Oaxaca), questioned as to the price of cochineal, prefaced his reply with the comment that it was "siempre muy variable." [65]

The cochineal registered in Oaxaca exceeded one million pounds on ten occasions between 1760 and 1782, but never thereafter. Output declined, albeit irregularly, over the closing years of the eighteenth century and the first two decades of the nineteenth century, with notable troughs in 1805, 1812, and 1813. Humboldt (1803–1804) observed that "in general the nopaleries increase very slowly in the Misteca." [66] After the Napoleonic wars the recovery of the industry was equally erratic; but by *ca.* 1850, the quantity of cochineal registered in Oaxaca again approached one million pounds.[67]

Central America

In the opening decades of the seventeenth century, attempts were made to promote the rearing of *grana fina* in the peninsula of Yucatán. The governor, Antonio de Figueroa, ordered that *arboles de calidad* should be planted; and he also brought Indians from Tlaxcala to give instruction on the cultivation and preparation of cochineal.[68] These efforts were only partly successful. The *memorial* of 1620 refers to "una nueva grana que llaman de Campeche," cultivated since 1615, and yielding about 1,500 *arrobas*.[69] However at least twice as much of the "new *grana*" was needed to obtain the results obtained from that raised in Tlaxcala or the Mixteca. It would appear that the product was chiefly *silvestre* or, at best, adulterated *fina*. The industry decayed in the later seventeenth century; in northern Yucatán demands for cochineal, among other products, under the *repartimiento* system prompted the migration of Indians to the south.[70] An *informe* of 1765–1776 states that good cochineal was plentiful and that cultivation might profitably be encouraged.[71] Commercial production appears to have been confined to the *municipio* of Chemax, whence quantities of *grana* were sold to the leather workers (*talaberteros*) of the province.[72]

The *memorial* also includes a brief and unflattering description of the cochineal of *Chiapas*.[73] The "cultivation" of *silvestre*, "unknown before the conquest," had commenced thirty or forty years earlier. Apparently, the insects were harvested from severed joints of nopal attached to the branches of trees; the joints withered or rotted, and the *grana* itself was of poor quality—one-third the value of *fina*. Other authorities took a different view. According to Vázquez (*ca.* 1612), the Indians of Chiapas "gathered quantities of fine cultivated cochineal, and wild from the mountains." [74] Thomas Gage (*ca.* 1630) wrote enthusiastically about the province of Zoques,

> the richest part of Chiapa . . . the chief commodities are silk and cochineal; whereof the latter is held the best of America, and the store of it so great that no one province alone exceeds it. Few Indians there are who have not their orchards planted with the trees whereon the worms breed . . . ; not that the Indians themselves esteem otherwise of it than as they see the Spaniards greedy after it, offering them money for it, and forcing them to the preservation of it in those parts which proved most successful for this kind.[75]

Gage had considerable first-hand knowledge of Central America, having spent six months in Chiapas and about ten years in Guatemala (1627–1637); however the statement that the cochineal of Chiapas was of the highest quality and equaled in quantity that of any other province was not merely an exaggeration but obviously quite untrue. Little more is heard of the industry over the next one hundred years. From the early 1740's attempts were made to foment cultivation in the territory of Zoques, but with meager results;

---

63 Latorre (ed), 1920: p. 85 (Socopo, Yucatán).

64 Hamnet, 1971*a*: p. 18; 1971*b*; p. 62.

65 Vasco (1776), 1963: pp. 46–47; similarly, Magán (1776), 1963: p. 74.

66 Humboldt, 1811: **3**: p. 64.

67 Sartorius ([1850], 1961: p. 172) put the amount exported from Mexico as a whole at between one and 1.5 million lbs.

68 Dahlgren de Jordan, 1961: p. 394 (unattributed quotation).

69 *Memorial* (1620), 1931; pp. 501–502. *Cf.* Chaunu and Chaunu, 1955–1959: **6**: p. 981 (cochineal in Seville from Honduras and Campeche, notably in 1616 [253 *arrobas*]). *Le tarif général de France,* 1664, distinguished between *cochenille mistèque, cochenille campeschiane,* and *cochenille sylvestre commune* (*Encyclopédie methodique,* 1782–1830: **67** [*Commerce,* 1783, I]: p. 531). Pomet in his *Droguier curieux* (1709: p. 13) gives *cochenille mestec, cochenille campeschane,* and *cochenille silvestre.*

70 Scholes and Roys, 1948: p. 305.

71 Rubio Mañe (ed.), 1938: pp. 46–47.

72 Dahlgren de Jordan, 1961: p. 394 (unattributed quotation). This is the only known reference from the New World to the use of cochineal for staining leather. The practice may have been introduced from southern Spain. At the time, cattle were important in the economy of Yucatán.

73 *Memorial* (1620), 1931: p. 500.

74 Vázquez de Espinosa, 1942: pp. 205 ("Criasse cantidad de Grana fina que se beneficia . . ."), 207 (province of Zendales).

75 Gage (1648), 1928: p. 167; see also pp. 152, 159.

output was about 700 *arrobas* in 1770.[76] Antonio de Alcedo (1786–1789) referred to only the wild product ("grana . . . sin beneficio"), used for dyeing cotton.[77]

Cochineal harvested in Guatemala (and in territories farther south) before the early nineteenth century was chiefly *silvestre*. From time to time efforts were made to put the industry on a surer and more profitable footing. In 1617 the Conde de la Gómara, governor of the province, turned his attention to parts of the southwestern highlands.[78] Many nopalries were planted, others extended, and expectations ran high. Vázquez (1620–1621) claimed that "in [Guatemala] . . . they gather quantities of fine grain or cochineal ("grana fina o cochinilla"); [the] speciality has made, and is making, great progress. . . ." The city of Guatemala attracted merchants "with silver and merchandise to exchange for cacao, indigo and cochineal . . . which the country offers in abundance." [79] Nevertheless Gómara's particular enterprise collapsed within a decade or two, probably for a variety of reasons. The areas chosen were far from ideal, being subject to heavy winter rain; severe plagues of locusts, notably in 1618, may have decimated many new nopalries; and, most important of all, Indian labor, as well as being scarce (indigo plantations made large demands), was also unskilled in the arts of cultivating and preparing fine cochineal. Nor was the time exactly opportune, for competition was growing as output of *grana* increased in the long-established centers of southern Mexico.

The most convenient shipping point for the products of Guatemala and of neighboring parts of Chiapas was Caballos in Honduras. John Chilton (1570) first referred to the export of cochineal "although it be not of such value as that of Nova Hispania." [80] Later, Dutch piracy in the Gulf of Honduras (Puerto de Caballos was virtually abandoned between 1625 and 1642) forced traders to carry their *grana* to Nicaragua, whence it was exported *via* the Laguna Grande and San Juan to Cartagena or Porto Bello (Panama).[81] Cochineal was also raised in highland Nicaragua [82] and in parts of Honduras and Salvador. "In the year [1636–1637] that I was in Granada," wrote Thomas Gage, "in one day there entered six *requas* (which were at least 300 mules) from St Salvador and Comayagua, laden with . . . indigo, cochineal and hides." [83] About one hundred years later, John Cockburn reported from southern Honduras—Salvador: "Here is abundance of cochineal, with which [the Indians] frequently pay their tribute. . . ." [84] But on the whole the cochineal industry of Central America declined in importance from the middle of the seventeenth century and, in terms of international trade, it had virtually disappeared by 1800.

POST-COLONIAL

Important changes occurred in the distribution and organization of cochineal production in the decades immediately following the fall of the Spanish empire in the Americas. By mid-century Guatemala had surpassed Oaxaca as the leading source of supply, only to be overtaken, in turn, by the Canary Islands. Significant quantities of cochineal from Peru, India, Java, and the lands around the western Mediterranean also entered world trade.

Attempts were made to extend the cultivation of *grana fina* from southern Mexico to parts of Guatemala (chiefly Verapaz) at the very end of the colonial period (1810–1820).[85] The world market was then depressed and the enterprise, organized on traditional lines, completely failed. A new start was made, under more propitious circumstances, in southwestern Guatemala soon after 1820 (map 7). The chief centers, Antigua Guatemala and Amatitlán, about twenty kilometers apart, developed an association based on the cycle of production. *Semilla* from the more tropical Amatitlán was regularly used to stock nopalries in Antigua. Other places where *grana* was raised before *ca.* 1850 included Chiquimula and Salamá in the same general area. For the first time in the New World production was organized primarily in terms of estates (*fincas*) owned or rented by Spaniards or *mestizos* and worked with free Indian labor. E. G. Squier observed (1858) that around Antigua "most of the cochineal lands belong to the municipality, and are let on leases of nine years duration" [86] (i.e. the normal life of the plantation). A few holdings amounted to three hundred or more acres, but such were "not nearly as productive as those of smaller size (commonly two to twenty acres), as the enormous number of people who must be employed to work them causes a confusion and great loss of labour." [87] The industry prospered for a

[76] Dahlgren de Jordan, 1961: p. 395.

[77] A. Alcedo (1786–1789), 1967: 1: p. 291; 4: p. 296. *Cf.* Campbell (1741), 1972: p. 135; Alzate (1777–1794), 1852: p. 86.

[78] Macleod, 1973: pp. 172–173.

[79] Vázquez de Espinosa, 1942: pp. 216, 237.

[80] Chilton, 1904: p. 367. See also D. Alcedo (*ca.* 1780), 1883: p. 451 (? *ca.* 1600).

[81] Vázquez de Espinosa (*ca.* 1628), 1942: p. 249; Gage (1648), 1928: p. 342.

[82] Vázquez de Espinosa (*ca.* 1613), 1942: p. 260; Dahlgren de Jordan, 1961: p. 395; Macleod, 1973; p. 172. Sloane's hearsay account (1707–1725: 1: p. lxxviii) of cochineal culture on the "Miskito" coast is dismissed as a mistake by Conzemius (1932: p. 64).

[83] Gage (1648), 1928: pp. 342–343.

[84] Cockburn (1730), 1810: p. 37. Squier (1870: p. 142) observed that "the nopal is abundant and indigenous in the plain of Comayagua, where its leaves are silvered with the webs of the *Cochineal silvéstre*" (then apparently unexploited).

[85] Squier, 1858: p. 519; Dahlgren de Jordan, 1961: pp. 395, 397; Hamnet, 1971*a*: p. 144.

[86] Squier, 1858: p. 507.

[87] Dunlop, 1847: p. 125. Stephens ([1839–1840], 1846: 2: pp. 277, 294) referred to cochineal plantations or *haciendas*.

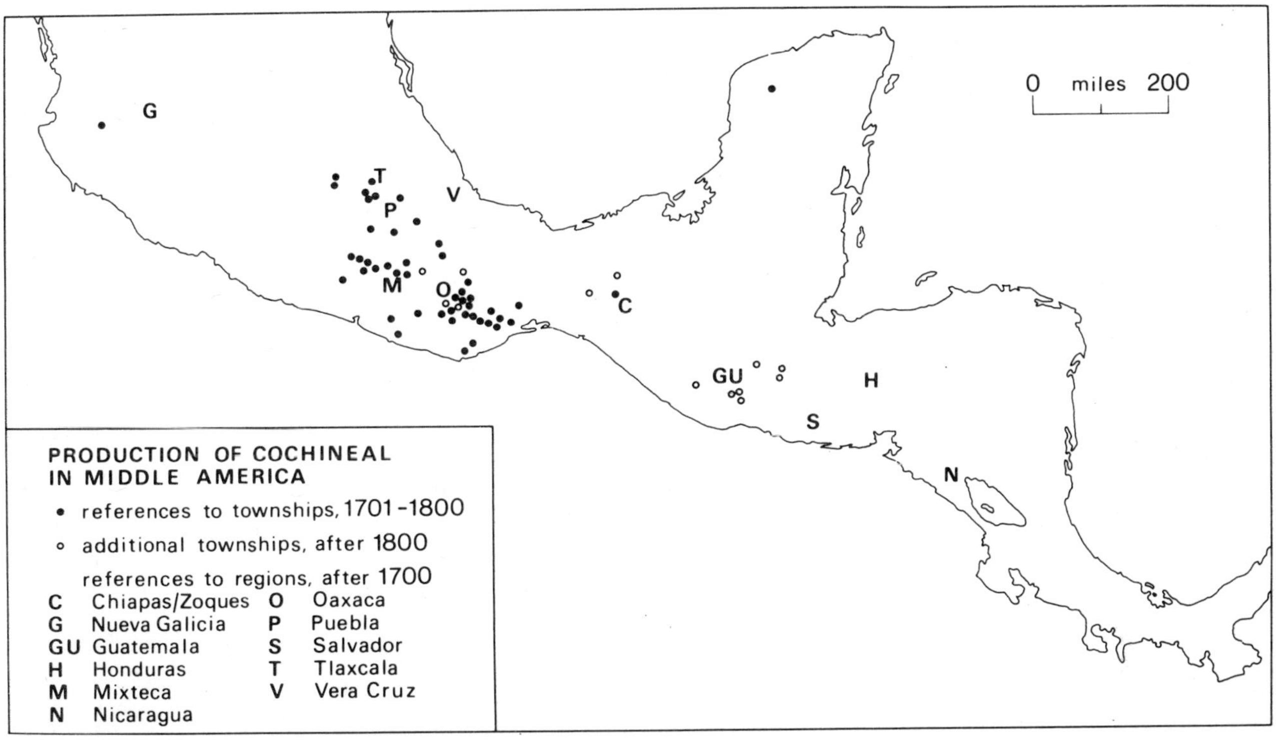

MAP 7. Cochineal in Middle America (1701 to 1800+).

number of reasons. As might be expected in a new area, both the nopal and the cochineal insect were relatively free from pests and diseases, particularly in Antigua.[88] Under the influence of local *sociedades económicas* (founded in the late eighteenth century) certain innovations were introduced, such as *galeras de adobe* (or *almacenes*) for reserving the *semilla* during the wet season, and an improved kind of oven for killing and drying the insects.[89] The industry was also exempted from taxation,[90] while controls on the quality of the product effectively prevented adulteration. Above all, perhaps, the trade in *grana* was not monopolized by Spain at a time of growing demand. Foreign merchants were active and supplies were shipped directly to the various European markets. Before 1840, cochineal had become "one of the great staple commodities of [Guatemala]," gradually supplanting indigo (*añil*) in first place.[91]

The industry was especially prosperous in the neighborhood of Amatitlán, notwithstanding the relatively high cost of labor. R. G. Dunlop (1844–1847) found that "nearly all the cultivators [in Amatitlán] are well off, and many who were without means a few years ago are now rich for Central America." [92] Arthur de Morelet (1857) wrote of the town "surrounded by its broad cactus fields, glistening like silver under their precious coating of cochineal. . . ." [93] Exports from Guatemala rose from 15,000 lbs in 1827 to 45,000 lbs in 1830, 105,000 lbs the following year, and then to a first peak of nearly 1.5 million lbs in 1849. A trough in 1853 (about 313,000 lbs) was immediately followed by the largest total of all, over 1.75 million lbs.[94] Signs of decline appeared in the 1860's when competition for labor from the coffee *fincas* began to be felt. Soon afterwards, direct competition from the new dyestuffs fatally crippled the industry, not only in Guatemala [95] but also in Oaxaca and in the Old World where alternative employment was not so readily available.

From about 1875 there are comparatively few reports of the rearing of cochineal; rather we learn of abandoned nopalries and of poverty and distress in former producing areas. A survey of the state of Oaxaca in 1891 emphasized that output had greatly

[88] Dunlop, 1847: p. 130; Dahlgren de Jordan, 1961: p. 398.
[89] Dunlop, 1847: p. 134.
[90] Haefkens (1827), 1969*a*: p. 105; (1832), 1969*b*: p. 306.
[91] Montgomery, 1838–1839: p. 132; Rubio Sánchez, 1961: p. 18.

[92] Dunlop, 1847: pp. 131, 135. The population of Amatitlán grew from *ca.* 3,000 to *ca.* 15,000 between 1830 and *ca.* 1858 (Squier, 1858: p. 507), the main period of expansion of the cochineal industry.
[93] Morelet, 1871: p. 412.
[94] Squier, 1858: p. 520; similarly Morelet (1857), 1871: p. 428. Before *ca.* 1830 total production probably exceeded exports by a considerable margin (Dunn, 1829: p. 231).
[95] Exports dropped precipitously to 18,401 lbs in 1883, and to 812 lbs in 1884 (Jones, 1966: p. 385).

declined, and *grana* was mentioned under only six townships.[96] Here and elsewhere in Middle America the production of cochineal continued into the twentieth century, less to meet the small world demand (increasingly supplied by Peru) than to satisfy local or regional needs. *Grana* from Antigua or Jocotenango was employed, mixed with other dyestuffs, in the blanket workshops of Momostenango, Guatemala, in the 1940's.[97] It could also be purchased in some of the markets of Yucatán.[98]

3. SOUTH AMERICA

PRE-HISPANIC

The textiles of prehistoric Peru were among the most remarkable achievements of the aboriginal New World.[1] Finely woven and elaborately patterned, they have survived for our inspection chiefly at burial sites in valleys along the arid coastal plain. Only in western South America were animal fibers employed: wool of the alpaca and occasionally of the llama and of the two related wild species, the vicuña and the guanaco. The record begins in the Early Formative period (*ca.* 700–500 B.C.). During the Late Formative (Paracas Necropolis) superb examples of the weaver's art suddenly make their appearance, to be followed, in the ensuing Classic and Post Classic, by variations on the same tradition. In the final, Inca, period fine cloth was highly prized, demanded as tribute and stockpiled as a form of wealth.[2] Pizarro's secretary, Francisco de Xérez, wrote of Cajamarca (1534): "The cloth was the best that had been seen in the Indies. The greater part of it is of very fine wool, and the rest of cotton of rich colors, beautifully variegated." And again: "It is wonderful how highly this cloth is prized in Spain for its workmanship. It is looked upon more as silk than as wool . . . [with] many patterns . . . very well embroidered." [3] Pedro de Cieza de León (1550) and Bartolemé de Las Casas (1559) also refer to striking colors,[4] combinations of shades of yellow, blue (indigo), and red, as well as the natural color of the fiber. Red was particularly prominent,[5] and the dyes employed have been the subject of considerable discussion.

G. A. Fester has identified cochineal (*Dactylopius coccus*) in textiles from the Nazca (Classic) and Chimu (Post Classic) periods.[6] In earlier publications the same author argued for the use of *Dactylopius* sp. in fabrics from Paracas (*ca.* 500 B.C.–A.D. 250);[7] but subsequent work showed that the principal red was madder, obtained from *Relbunium* sp. (relative of *Rubia* of the Old World). *Relbunium* has not so far been identified in designs of later date.[8] Fester maintained, on the basis of early colonial reports, that *D. coccus* "must have been domesticated widely during the Inca period." Wild species (*D. confusus, D. ceylonicus*) were probably also collected. Red dye in a *manta* from a Bolivian tomb ("de edad probablemente no muy elevada") proved on examination to be of *D. coccus*.[9] Another tomb, this time in the Valle de Calchaquí (Salta, northwest Argentina) contained "grana de tunilla ó sea restos de cochinilla indígena," [10] probably *D. ceylonicus*. Some of the Peruvian textiles portray *cardones* (*Opuntia* spp.),[11] and combs and pins of *O. exaltata* (host of *D. confusus*) have been found in burial sites along the coast.[12]

Full appreciation of the importance of cochineal in pre-Columbian times must await further chemical and spectroscopic work on fabrics of known date and provenance. Possibly *Dactylopius* sp. became the leading source of red sometime in the Classic, which raises the question, discussed below, of a connection between the South and Middle American centers of production. Cochineal may also have been used in the preparation of lacquers and paints (some textiles were painted),[13] but the evidence so far available points more to inorganic compounds and, apparently, to *Bixa orellana*.[14]

COLONIAL AND POST-COLONIAL

SIXTEENTH CENTURY

Early references to *grana* and to words for cochineal in the principal aboriginal languages lend support to the view that the product was widely known before the Spanish conquest. The statements are generally

---

[96] Velasco, 1891: pp. 269, 291, 313, 320, 332, 344.

[97] McBryde, 1945: p. 142; O'Neale, 1945: p. 29. See also Osborne, 1935: p. 53; Start, 1963: p. 18 (quoting E. MacDougall): "Red is made from the cochineal insect; the dried bodies are ground on a mealey stone (*metate*), small limes and a little stannous chloride added and boiled in an earthenware pot."

[98] Steggerda, 1941: p. 148.

[1] Barnett, 1909: pp. 265–268; Crawford, 1916*a*: pp. 107–191; 1916*b*: pp. 417–431; Harcourt (1924), 1962; O'Neale and Kroeber, 1930: pp. 23–56; Means, 1932; O'Neale, 1949: pp. 123–126. There are excellent illustrations in Schmidt, 1929.

[2] Murra, 1962: pp. 710–728.

[3] Xérez, 1853: pp. 326, 334, 1872: pp. 31, 60. He clearly alludes to fabrics of alpaca or vicuña, the latter finest of all animal fibers.

[4] Cieza de León, 1864: p. 405; Las Casas, 1958: **1**: p. 217.

[5] O'Neale, 1937: p. 136.

[6] Fester and Lexow, 1943*a*: p. 91; Fester, 1954: p. 241. See also Valette, 1913: pp. 43–45; Cook, 1925: p. 104; Harcourt (1924), 1962: p. 6. Saltzman, Keay and Christensen (1963: p. 245) reported the presence of cochineal in twelve samples of fiber from Peru (Nazca; Tiahuanacu [*ca.* A.D. 1000]; and Chancay IV [A.D. 1450–1550]).

[7] Fester and Cruellas, 1934: p. 156; Fester, 1940: p. 233.

[8] Fester, 1954: p. 240.

[9] Fester, 1941: p. 102.

[10] Ambrosetti, 1908: p. 242.

[11] Mead, 1916: no. 28; Levillier, 1928: p. 17 (Paracas); Yacovleff and Herrera, 1934*a*: pp. 320–321 (Mochica, Nazca).

[12] Yacovleff and Herrera, 1934*a*: p. 318.

[13] Holmes, 1889: p. 17; O'Neale, 1949: p. 123.

[14] Nordenskiöld, 1931: pp. 95–100; Fester, 1940: p. 238.

brief and they do not, of course, serve to identify the species of *Dactylopius* involved; nor is it always clear to what extent the insect and the host cactus were cultivated. Indeed, as in Mexico, it is not even certain that references to the presence of cochineal imply that some collecting was done, although the probability of this seems strong.

In Quechua and Aymara cochineal is (or was) known as *macno* or some cognate word.[15] González Holguín (1586) described *macnu* (Quechua) as "color colorado par teñir."[16] The Quechua vocabulary of Domingo de Santo Tomas (1560) refers only to "grana color—*pucca*" (red).[17] The Chibcha for "colorado" was *heesen mague*.[18] Some of the chroniclers are more specific. Baltasar Ramírez (1597), writing of the province of Condesuyo (northeast of Arequipa), mentions "grana que alla se llama *magno*."[19] Martin de Murúa (*ca.* 1600) states that the *indios serranos* dyed alpaca wool "con *magno*, ques finísima grana."[20] The *Relaciones Geográficas* (1580's) refer to *magno* in the vicinity of Nazca and in the province of Rucanas Antamarcas (map 8, Appendix B); later reports too are confined to the southern highlands (Cuzco, Parinacochas, Aymaraes, Arequipa, Chumbivilcas). The word may therefore have had closer regional associations than has so far been recognized.

Macno meant not only cochineal itself but also the cakes of dyestuff (*tortas, pastillas, panecillos, pastas*): thus "gusanos de que hacen unos panecillos que los llaman los indios *magno*" (1586).[21] According to the botanist Hipólito Ruiz (1777–1788), cochineal and liver (*higado*), pounded together, dried in the sun and made into *tortas*, was known as *pircay* or *pilcay* in the area around Tarma, central Peru.[22]

Grana was first reported from the north coast of Peru in the early years of the Spanish conquest. The *relación* of Juan de Sámanos (based on the work of Francisco de Xérez, 1527)[23] refers to

> muchas mantas de la lana y de algodón, y camisas . . . y otras muchas ropas, todo lo mas de ello muy labrado de labores muy ricos de colores de grana y carmisi, y azul y amarillo, y de todas otras colores de diversas maneras de labores y figuras de aves y animales y pescadas y arboledas. . . .[24]

Miguel de Estete (1534), like Xérez a companion of Pizarro, observed that the Indians raised "mucha grana y algodón."[25]

The *Relaciones Geográficas*, in addition to providing evidence for the southern coast and highlands, extend the distribution of *grana* or *cochinilla* to Ayacucho and to regions farther afield—the highlands of southern Ecuador (Loja, Cuenca), central Bolivia (Sucre), northwestern Argentina (Tucumán), and Paraguay (Talavera). López de Velasco (1571–1574) relates that *cochinilla* was available in "part of the province of Río de la Plata" (probably Paraguay).[26]

Several of the early reports refer to the raising (*criar*) of *grana fina* or *cochinilla finísima*. Around Huamanga (Ayacucho) in 1586 the host plants (*cardones*) were carefully cultivated for the benefit of the insects (*gorgojitos*).[27] Some improvement may have been achieved at different times and places, only to disappear as interest declined. *Dactylopius coccus* was known in pre-Columbian Peru. *D. confusus*, today collected from hedges of *Opuntia exaltata* in the Cuzco-Puno area,[28] was also of greater importance in the past. In northwestern Argentina the species harvested were *D. ceylonicus* and possibly the inferior *D. tomentosus*, but at no time, according to Fester, was *D. coccus* cultivated.[29]

The question of a possible connection between the exploitation of cochineal insects in Middle America on the one hand and in South America on the other cannot yet be resolved. The evidence is not only fragmentary and partly indirect but necessarily different in kind between the two regions. It would therefore be disingenuous to stress the much earlier dates from South America. However, the belief that the exploitation and domestication of *Dactylopius* sp. first began in Middle America may be questioned; and the further suggestion that the Spaniards introduced cochineal to Peru, or at least played a significant part in its dissemination through western and southern South America, must

15 Cook (1925: p. 104) observed that the word was still current in some Indian markets.

16 *Cf.* Bertonio (1612): *makhno* (Aymara)—"tortilla de yerua con que tiñen lanas de colorado."

17 Nor is *macno* given by Fernando de la Carrera (1644).

18 Uricoechea, 1871. Arango Bueno (1963: p. 29) claims that cochineal was used by the Chibcha of Colombia; *cf.* Joaquín Acosta, 1848: p. 207. The Mayan word for cochineal was *mukay* (Roys, 1931: p. 337), which may be related.

19 Ramírez, 1936: p. 41.

20 Murúa, 1962: 2: p. 154.

21 Jiménez de la Espada (ed), 1965: 1: p. 234 (28). Middendorf (1890) translates *majnu* (Quechua) as "la grana fina, pasta de cochinilla fina." Grana pastilla = *macnu patara* in Aymara (Fester, 1940: p. 232). See also Ruiz (1777–1788), 1952: 1: p. 84: grana pasta = *maccnum* (Cuzco), *macano* (Chile).

22 *Ibid.*, p. 122. *Cf.* Bueno (*ca.* 1763–1780), 1951: p. 32; Haenke (*ca.* 1800), 1901: p. 173.

23 Porras Barrenechea (ed.), 1937: p. 66.

24 Sámanos, 1844: p. 197.

25 Estete, 1924: p. 49.

26 López de Velasco, 1894: p. 555.

27 Jiménez de la Espada (ed), 1965: 1: p. 192 ("Hay unos cardones . . . los cuales se cultivan con mucho cuidado; su fruto es gana colorado finísima con que se tiñe la ropa . . . ; y cuando ha de dar su fruto el cardón, ques unos gorgojitos muy pequeños, vivos y muy colorados, tienen los señores destas heredades, que son indios, gran cuidado de juntar todo género de huesos que noche y día estén ardiendo, con cuyo humo y calor dicen se crían los dichos gorgojuelos, ques la dicha grana.").

28 Fester, 1940: p. 232; 1941: p. 101; 1954: p. 241; F. Herrera, 1941: p. 313. *O. exaltata* is also found in Ecuador and Bolivia and probably northern Chile (Britton and Rose [1919–1922], 1963: 1: p. 76).

29 Fester and Lexow, 1943*a*: p. 92; Fester, 1954: p. 243.

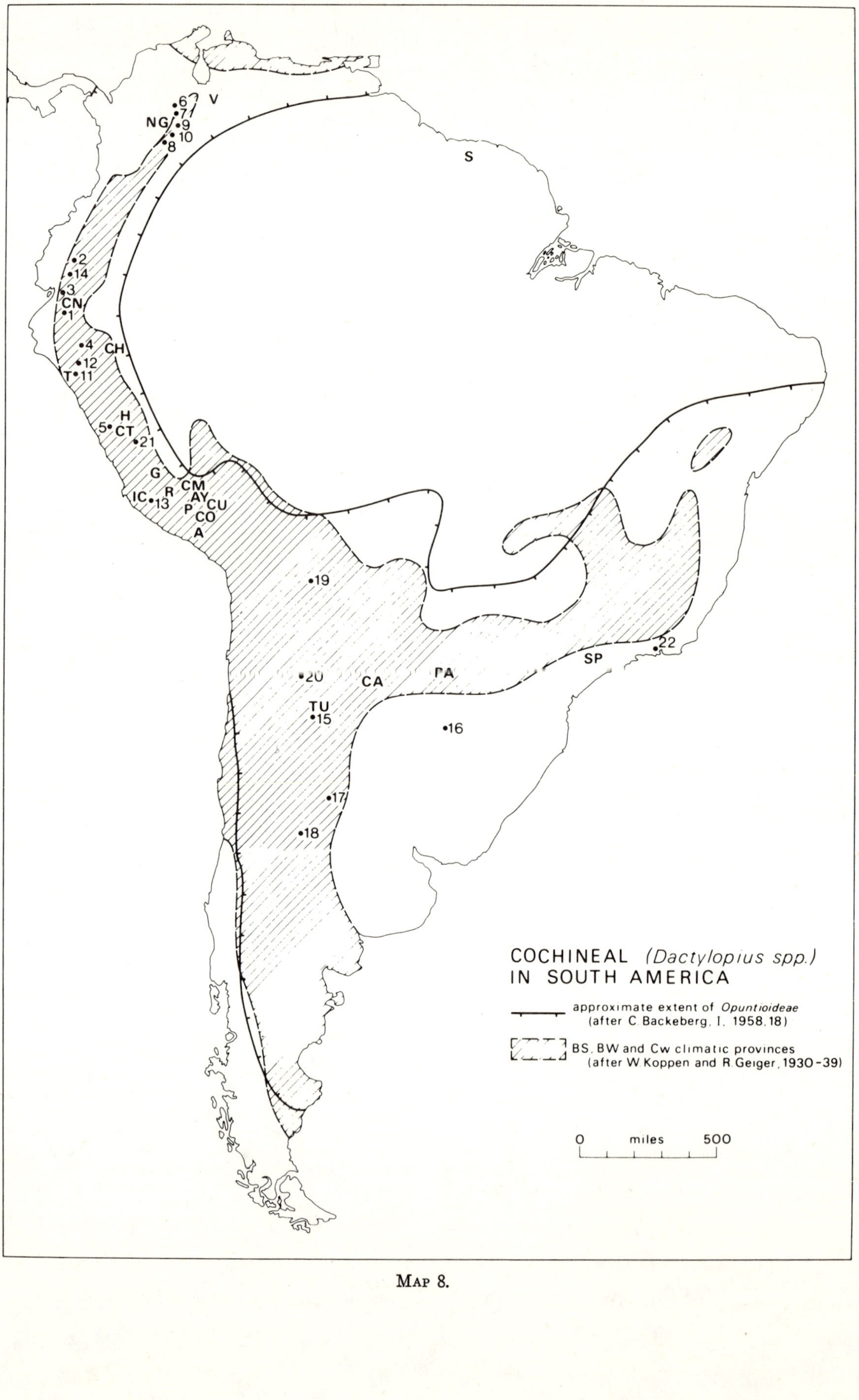
COCHINEAL (Dactylopius spp.) IN SOUTH AMERICA
approximate extent of Opuntioideae (after C. Backeberg, I, 1958, 18)
BS, BW and Cw climatic provinces (after W. Koppen and R. Geiger, 1930-39)
0 miles 500
V
NG
S
CN
CH
T
H
CT
G
CM
AY
CU
R
IC
P
CO
A
CA
PA
SP
TU
1
2
3
4
5
6
7
8
9
10
11
12
13
14
15
16
17
18
19
20
21
22

Map 8.

MAP. 8. *Dactylopius* spp. in South America (Appendix B)

Regions or provinces

A Arequipa
AY Aymaraes
CA El Chaco
CH Chillaos or Lulaychillaos
CM Chumpivilcas or Chumbivilcas
CN Cuenca
CO Condesuyo
CT Cajatambo
CU Cuzco
G Guamanga or Huamanga (Ayacucho)
H Huanuco
I Ica
NG New Granada
P Parinacochas
PA Paraguay
R Rucanas Antamarcas
S Surinam (c)
SP São Paulo
T Trujillo
TU Tucumán
V Venezuela

Environs of towns and cities

[numbers have no significance other than identification]

1 Loja or Loxa
2 Ambato
3 Cuenca
4 Jaen de Bracamoros
5 Cajatambo
6 Cúcuta
7 Pamplona
8 Tunja
9 Soacha
10 Sogomoso
11 Conaché
12 Cajamarca
13 Nazca
14 Guananda
15 Tucumán
16 Talavera
17 Córdoba
18 San Luis
19 La Plata (Sucre)
20 Valle de Calchaquí
21 Tarma
22 Rio de Janeiro

be rejected. There is nothing to support such a view; on the contrary, it seems that the insect was already widely known and valued as a source of dye at the beginning of the sixteenth century and that locally some progress had been made in the direction of domestication. Cochineal itself was not exacted as tribute by the Inca state, as it was in Mexico under Moctezuma, but cloth, some of it patterned in fast and brilliant colors, was regularly demanded. The Peruvians had available excellent animal fibers that took cochineal more successfully than cotton, the only important textile material in Middle America. It is therefore by no means inconceivable or even unlikely that knowledge of the dye and of the techniques employed in its preparation were carried in late prehistoric times along the western seaways from Peru or Ecuador to southern Mexico, perhaps the coast of Oaxaca or Michoacán (the shellfish-dye industry may have followed the same route). The most reasonable alternative hypothesis appears to be that of independent origins, rather than movement in the opposite direction.[30]

SEVENTEENTH AND LATER CENTURIES

Father Bernabé Cobo (1650) confirms that the *cochinilla* of Mexico was known as *magno* in Peru,[31] but it is not entirely clear that his striking description of the domesticated insects "as large as peas or beans" is meant to apply to the latter area. In the second half of the eighteenth century, cochineal was employed in the *obrajes* of Cajatambo. Apparently little or no attention was paid to the rearing of the insect;[32] on the other hand, the "plantiós de grana" in the province of Parinacochas may imply some improvement.[33] Hipólito Ruiz (1777–1788) observed that cochineal was produced but not cultivated in Huanuco, and that around Tarma in particular the full potential was not realized.[34] In the region of Arequipa (1765) the Indians sold *macno,* prepared from wild cochineal, to buy cloth (*bayetas*) and coca.[35] Vázquez de Espinosa (*ca.* 1620) tells us nothing about these areas but remarked that "the Indians [of La Plata, i.e. Sucre, Bolivia] sell wool dyed in different colors, and bricks of wild cochineal ('unos pañesillos de grana silvestre') with which they make varicolored dyes by combining it with various plants." [36]

The exploitation of cochineal was one of the chief industries of southern Ecuador in the eighteenth century. Loja in the 1740's obtained "a very great advantage from breeding the cochineal," and some regarded the dyestuff as similar in quality to that of Oaxaca.[37] Dionysio de Alcedo (1766) claimed only

[30] Discussions of pre-Columbian contact between northwest South America and Mesoamerica (see, in particular, Borhegyi, 1959: pp. 143–151; Evans and Meggers, 1966: pp. 243–264; Lathrap, 1966: pp. 265–276) do not, so far as is known, refer to cochineal; but there is general agreement that movement during the centuries immediately prior to the Spanish conquest was predominantly from south to north.

[31] Cobo, 1956: 1: p. 202.

[32] A. Alcedo (1786–1789), 1967: 1: p. 263; Bueno (*ca.* 1763–1780), 1951: p. 39 (1767); Haenke (*ca.* 1800), 1901: p. 206.

[33] A. Alcedo (1786–1789), 1967: 3: pp. 134, 296; Bueno (*ca.* 1763–1780), 1951: p. 77 (1767); Haenke (*ca.* 1800), 1901: p. 259.

[34] Ruiz, 1952: 1: pp. 83, 122, 130.

[35] Bueno (*ca.* 1763–1780), 1951: p. 85 (1767).

[36] Vázquez de Espinosa, 1942: p. 655. Harcourt (1962: p. 6) reported (1924) that the Indians of Cajamarca dyed black with cochineal, using a mordant.

[37] Ulloa (1748), 1760: 1: p. 341; similarly Caldas y Tenorio

that *grana silvestre* was produced in quantity and used in dyeing cloth and carpets.[38] The province of Cuenca supplied "abundancia de quina y cochinilla" (1786–1789).[39] Farther north, in the vicinity of Ambato, there was less demand and thus apparently less attention to both output and quality.[40] Caldas y Tenorio in his *Memoria sobre la importancia del cultivo de la cochinilla* (*ca.* 1810) noted the wide range of physical conditions under which the insect bred in Ecuador and New Granada.[41] The dyestuff was probably known in New Granada at the time of the Spanish conquest, as Caldas y Tenorio maintained. A description of Porto Bello, near Panama, dating from the early seventeenth century, states that ships arrived there from Lake Maracaibo "laden with indigo and cochineal." [42] John Campbell (1741) included cochineal among the products of the province of Santa Marta.[43]

The distribution of *Dactylopius* spp. extended through the valleys and intermontane basins of southern Peru and Bolivia to northwestern Argentina. In the town of Londres (Tucumán), about 1620, cloth and blankets were dyed with local cochineal.[44] Apparently the color was satisfactory although there was no appreciable cultivation.[45] At the end of the eighteenth century, the *grana* of Tucumán was considered to be of little value.[46] In areas farther south the wild insects were harvested. By this means the Indian *tributarios* of Córdoba paid their taxes, and around San Luis the dyestuff was sold by "la jente pobre de la campaña" in order to purchase cloth (*lencería*) from Chile.[47] An early reference (1580's) to *grana* in northern Paraguay has already been mentioned. There is a Guarani term for the dye [48] which was prepared by the Chaco Indians into modern times.[49]

(1810), 1912: p. 432 (". . . las muestras que poseemos de la de Loja no ceden á la más bella de Méjico").

38 D. Alcedo, 1915: p. 45.

39 A. Alcedo, 1967: **1**: p. 405.

40 *Ibid.*: p. 50; Ulloa (1748), 1760: **1**: p. 341.

41 Caldas y Tenorio (1810), 1912*a*: pp. 430–432. Humboldt (1811: **3** p. 65) refers to the wild cochineal of New Granada amongst other areas.

42 Lewin (ed), 1958: p. 121. Chaunu and Chaunu (1955–1959: **6**: p. 981) record small shipments of cochineal (1610, 1616) from *tierra firme,* possibly Central America. *O. Ficus-indica, O. monacantha,* and *Nopalea cochenillifera,* all cochineal hosts, are known from Colombia (Pérez Arbelaez, 1947: p. 393). According to Hall (1824: p. 31), the insect was "found abundantly" near Timotes and Barquisimeto in the Andes to the south and east of Lake Maracaibo.

43 Campbell, 1972: p. 195. See also Spence, 1878: **1**: p. 53 (Venezuela).

44 Vázquez de Espinosa, 1942: p. 676.

45 León Pinelo (1650), 1943: pp. 248–249.

46 A. Alcedo (1786–1789), 1967: **4**: p. 118. Andrews (1827: **1**: p. 184) lists cochineal among the products of the province of Tucumán.

47 Quesada (ed), 1965: p. 564 (1788).

48 Ruiz de Montoya (1639), 1876.

49 Grubb, 1904: p. 83; 1911: p. 67; Métraux, 1946: p. 289.

Early in the nineteenth century the cochineal insect flourished "in many parts of the province of São Paulo," [50] but whether it was also cultivated is not known. Some decades before, a small plantation for the raising of cochineal was started near Rio de Janeiro,[51] and it was probably from here that insects were first taken to India. When John Barrow called at Rio in 1792 he examined a building in the *Paseo Publico,* the interior walls of which were decorated with paintings "descriptive of what was then considered the eight most important productions of the Brazils." One showed "a plantation of the Cactus Opuntia, with the mode of preparing cochinille from the insect." But, he added, "the quantity raised at present is exceedingly trifling." Another garden, belonging to the crown, was "chiefly intended for the cultivation of that species of cactus on which the cochinille insect feeds, and for the reception of curious or interesting plants; but we found it greatly neglected, and the collection very limited. . . ." [52] Prince Maximilian of Wied-Neuwied reported (1815–1817) that "formerly a considerable quantity of cochineal was raised [in Sagoarema, about forty kilometers east of Rio de Janeiro], but its cultivation has been relinquished. It was purchased by the king at the rate of half a *doble* (6,400 *reas*) a pound; but the planters themselves destroyed this lucrative trade by adulterating . . . with flour." [53]

Grana was collected to meet local needs in many parts of South America during the nineteenth century. Commercial production was much more restricted. James Orton found (1870) that "the *Opuntia tuna* [was] extensively cultivated" for cochineal in the southern part of the basin of Quito.[54] In Peru about the same time the industry was most prominent around Trujillo, Ayacucho, Arequipa, and Ica. The province of Ica is said to have produced 30–40,000 lbs of cochineal annually.[55] According to T. J. Hutchinson (1873), the output of a single plantation at Conaché, northeast of Trujillo, was over 18,000 lbs.[56] In the second half of the century Peru became one of the chief exporters of cochineal and today meets a large part of the world demand. The industry is concentrated around the city of Ayacucho. Not much attention is given to cultivation or improvement, and col-

50 Spix and Martius, 1924: **2**: p. 19. See also Diderot and d'Alembert, 1751–1757: **3**: p. 559; Fester and Lexow, 1943*a*: p. 91.

51 Staunton, 1797: **2**: pp. 164, 169.

52 Barrow, 1806: pp. 81–83; similarly Abel (1816), 1818: p. 19.

53 Maximilian of Wied-Neuwied, 1820: p. 56; *cf.* Burton, 1869: **1**: p. 94. Some Brazilian cochineal was imported through Lisbon (*Reports,* 1839: p. 26).

54 Orton, 1870: p. 108. See also Hall, 1824: p. 31.

55 Paz Soldan, 1862: p. 566; also pp. 211 (Trujillo), 362 (Ayacucho). *Cf.* Markham, 1892: p. 493 (Arequipa).

56 Hutchinson, **2**: p. 147.

lecting is mostly a part-time activity; nevertheless, as much as 135,000 kilograms was exported in 1966.[57]

4. THE OLD WORLD

TRADE IN COCHINEAL

EUROPE

The date of the first consignment of cochineal to Europe is unknown. The years 1523 [1] and 1526[2] have been suggested but without, apparently, much documentary support. León Pinelo (*ca.* 1650) states that shipments from New Spain commenced in 1543.[3] This appears to be somewhat late in view of the Spanish government's interest in the dyestuff and the favorable comparisons that were drawn with kermes. Cochineal proved to be ten or twelve times more powerful pound for pound, and it partly displaced the older product, at least in the main cloth-producing centers of western Europe, before the end of the sixteenth century. Most probably small quantities of American cochineal reached Europe in the 1520's and 1530's, to be tried experimentally in Spain's own cloth industry. One of several obstacles to any substantial export in the two decades immediately following the conquest was presumably that of supply, until then related to the rather specialized Indian demand. However, by 1554, according to Cervantes de Salazar, "a great abundance of [cochineal], at great cost, [was] imported by Spain." [4]

Cochineal became an increasingly important item in the export trade of New Spain in the second half of the sixteenth century, partly through the efforts of particular viceroys to increase production, partly because of the organization of annual *flotas* from the Indies after 1564, and partly in response to the growing demand in Europe. A dispatch from Spain on July 18, 1565, announced the arrival in Seville of one convoy carrying gold; another, described as "the cochineal fleet," was expected in the middle of August.[5] Intermittent figures of the quantities shipped are available from the 1570's: 7,000 *arrobas* (175,000 lbs) in 1575, 12,000 in 1576, and 9,200 in 1580.[6] José de Acosta (1590) records that 5,677 *arrobas,* worth 283,750 *pesos,* arrived in 1587.[7] Amounts also varied considerably in the final decade of the sixteenth century: 14,000 *arrobas* in 1591, 13,000 in 1594, 7,000 in 1598, and 6,000 in 1600.[8] Gómez de Cervantes (1599) put the annual figure at 10,000 to 12,000 *arrobas,*[9] probably a somewhat generous estimate. Cochineal stood in second place by value of all goods exported from the Spanish territories in the New World, but in 1594 this amounted to only 2.8 per cent of the total in comparison with precious metals at 95.6 per cent.[10] León Pinelo described *grana* as "el mas precioso [género] que se trae de las Indias, despues oro, plata y perlas, y viene en mucha cantidad." [11] The greater part was *grana fina* from Tlaxcala and Oaxaca; in 1620 quantities for the previous year were estimated at 6,000 to 6,500 *arrobas* of *fina,* 1,500 of *silvestre* and about the same amount of "una nueva grana que llaman de Campeche." [12]

These figures were greatly exceeded in the eighteenth century: approximately 35,000 *arrobas* in 1736 [13] and from 19,000 to 32,000 between 1760 and 1772.[14] England alone took 6,765 *arrobas* in 1773 and 4,168 in 1782, re-exporting up to a quarter in the former year.[15] In 1786 cochineal represented approximately 6 per cent by value of all American products arriving in Cádiz.[16] About 6,000 *arrobas* left Vera Cruz in 1796 and 16,000 in 1820; between the two dates amounts varied from 838 *arrobas* (1797) to over 43,000 *arrobas* (1802).[17] Guatemala became an independent exporter in the 1820's and in 1858 led the world with a total of about 80,200 *arrobas.*[18] Three years later the center of production shifted, for the first time, to the Old World, when the Canary Islands exported over 84,200 *arrobas.*

During the whole of the colonial period cochineal could legally be imported only through Cádiz or Seville, and no doubt the bulk of supplies reached Europe in this way. The Spanish-American trade was not organized by great companies, such as developed in

---

[57] Gade, 1972: p. 62 (25,000 kgs in 1932). Brand (1966: p. 84) gives 60,000 kgs in 1937, 47,000 kgs in 1951, and 82,000 kgs in 1962.

[1] Yeats, 1887: p. 385; Watt, 1889–1893: **2**: p. 399; 1908: p. 347; Diguet, 1928: p. 481. See also A. Herrera (*ca.* 1600), 1934–1956: **8**: p. 25; Chipman, 1967: p. 100.

[2] Gillespie, 1920: p. 133; Dahlgren de Jordan, 1961: p. 391.

[3] León Pinelo, 1943: **1**: p. 247.

[4] Cervantes de Salazar, 1953: p. 61.

[5] *State Papers* (1558–1571), 1916: p. 327.

[6] Lee, 1948: pp. 458–459; Chevalier, 1963: p. 193.

[7] José de Acosta, 1962: p. 183. In 1587 the Venetian ambassador to Spain provided the Doge and Senate with a list of goods that had arrived from New Spain; they included "5,600 cases of cochineal" (*State Papers* [1581–1591], 1894: p. 308); *ibid.* (1607–1610), 1904: p. 186, a similar communication (1608).

[8] Lee, 1948: pp. 459, 472; Chevalier, 1943: pp. 327–328. Pike (1966: p. 65) gives 16,000 *arrobas* for 1594.

[9] Gómez de Cervantes, 1944: p. 163. *Cf. Memorial* (1620), 1931: p. 501 (10–11,000 *arrobas*).

[10] Parry, 1966: p. 243. See also Chevalier, 1943: p. 327; Lee, 1948: pp. 462–463.

[11] León Pinelo (*ca.* 1650), 1943: **2**: p. 249.

[12] *Memorial,* 1931: p. 501.

[13] Diderot and d'Alembert, 1751–1757: **3**: p. 560.

[14] Hamnet, 1971*a*: p. 172.

[15] A. Anderson, 1787–1789: **4**: p. 450; p. 454, complete figures for 1773–1782. In 1777, goods shipped from England to St. Petersburg included cochineal (Coxe, 1784: **2**: p. 499). Milburn (1813: **2**: p. 209) put the average annual "consumption" by Great Britain between 1776 and 1800 at 171,169 lbs (i.e. about 6,846 *arrobas*). In 1832 McCullock (p. 283) estimated that 300,000 to 360,000 lbs were annually imported and a third or more of this re-exported.

[16] Hamnet, 1971*a*: p. 174.

[17] *Ibid.,* p. 175. *Cf.* Humboldt, 1811: **4**: p. 29.

[18] Born, 1938: p. 226. See also Dunlop, 1847: p. 135; Rubio Sánchez, 1961: p. 18.

England and Holland, but by individuals or more usually by partnerships between home-based merchants, supplying goods and capital, and representatives in the Indies, contributing chiefly their services.[19] Such forms of business contract (*commenda* and *societas*), based on profit sharing, were derived from Italian practices of the Middle Ages.[20] In Seville partnerships existed under the powerful merchant guild or *consulado*.[21]

Supplies of cochineal outstripped Iberian requirements by about 1560. Thenceforth an increasing proportion was re-exported, chiefly to the Low Countries, politically linked to Spain after 1535 and the center of the most thriving cloth industry. Antwerp became the leading market and entrepôt. Luigi Guicciardini, describing the trade of the port in the 1560's, refers to shipments of cochineal to Ancona, Venice, and Milan.[22] Other parts of Europe were also supplied, either directly from Seville or through the Netherlands (Amsterdam had displaced Antwerp by the end of the sixteenth century, and the price of Mexican *grana* was quoted on the exchange from 1589).[23] England was importing cochineal by 1558; records commence in 1569[24] and *quchinilla* "from southern Spain" is mentioned in a trade directory of about 1575–1585.[25] Foreigners resident in Seville—English, Flemish, French, and men from the leading Italian cities—dealt in cochineal and helped to finance operations.[26] The product often passed through a dozen or more hands between producer and consumer. Supplies were subject to many vicissitudes, including war and piracy,[27] and prices varied accordingly.[28] That large profits were made is certain, and the Spanish monopoly of production was a constant irritation to a country such as England with an expanding cloth industry but with little reputation for dyeing. Travelers to the Indies reported what they had seen of cochineal,[29] and the prospect of finding or "cultivating" the dyestuff, as well as other tropical products, prompted efforts towards exploration and colonization in the wake of the Spaniards. Thus Robert Harcourt wrote hopefully in 1613: "For if they in New Spaine and Peru have Cuchenille, Anir (indigo) and cotton wooll; we in Guiana have also . . . divers good commodities for Dyers, and likewise in all likelihood Cuchenille."[30] In 1636 the Company of Providence Island (Bahamas) instructed Captain W. Rous "to make strict inquiry where indigo, cochineal, sarsparilla, ginger, rice, any fruits, drugs, or other useful commodities may be found, fit to grow [on Santa Catalina]."[31] This suggests not only the importance attached to cochineal but also that those who sought it were not always well informed of what to look for.

ASIA

Cochineal reached the outposts of Asia soon after the middle of the sixteenth century. It was known and esteemed "hasta China y hasta Turquía, casi por todo el Mundo," according to Bernardino de Sahagún (*ca.* 1570).[32] Supplies passed along several trade routes: one from the Levant into Persia and thence to India; another from Constantinople and the ports of the Black Sea to Turkey and the Caspian region; a third directly to India in ships of the English East India Company; and a fourth from New Spain to eastern Asia via the Philippines. The dyestuff was well suited to long-distance commerce involving many intermediaries; it was also at all times costly and found a ready market only where luxury materials were manufactured, such as brocade and carpets in western Asia and silk in China. Cochineal was little used in dyeing Indian cottons, for which the indigenous lac was generally preferred.

Dyestuffs had long been an important item in the intercontinental trade of Venice. Kermes was still in demand in the 1560's;[33] but by this time the port also handled "cuchenille [in] great quantitie" that had been re-exported from Antwerp.[34] Its ultimate destination often lay far to the east. Venetian merchants had close ties with Constantinople, the ports of the Levant, and with cities throughout the vast hinterland of western Asia. By the Venice-Levant route "much cochineal" reached Aleppo, according to Pedro Teixeira (*ca.* 1605).[35] The traveler and company agent John Ander-

19 For an example involving cochineal, see Hakluyt, 1903–1905: **10**: pp. 166–167 (1590).

20 Sayous, 1929: pp. 282–301.

21 R. S. Smith, 1940: pp. 91–111.

22 Guicciardini, 1593: p. 34. The author is sometimes confused with his uncle, the historian Francisco Guicciardini (1482–1540); thus we find references to "before 1540" (Born, 1938: p. 217; *cf.* Brand, 1966: p. 82).

23 Posthumus, 1946: **1**: p. 420.

24 Lee, 1951: pp. 207–208.

25 Tawney and Power (ed), 1924: **3**: p. 202.

26 Girard, 1932: pp. 456–459 (seventeenth century); Pike, 1966: pp. 17, 68.

27 *State Papers* (1587–1603), 1899: p. 563; *ibid.* (1589–1590), 1964: pp. 177 (Dutch ship), 356 (Spanish ship).

28 Lee, 1951: pp. 200 ff.

29 Bodenham (1564), 1904: p. 360; Chilton (1568), 1904: pp. 362–364, 367; Hawks (1572), 1904: p. 380; Philips (1580–1582), 1904: p. 435.

30 Harcourt, 1928: p. 59. Linnaeus is said to have received cochineal insects from Surinam in 1756 (Ellis, 1763: p. 663; Olivier, 1791: p. 98; *Reports,* 1839: p. 14). In a letter to John Ellis dated December 8, 1758 (J. E. Smith, 1821: **1**: p. 111), he refers only to America (*cf.* Linnaeus, 1758: **1**: p. 457). Schomburgk (1848: **3**: p. 841) mentions *O. tuna* and *O. cochenillifera,* but not cochineal. The *coccinella* of Merian (1719, p. 2) does not appear to have been *D. coccus.* W. E. Roth's ethnographic accounts of Guiana (1915, 1924, 1929) are silent on the matter.

31 *State Papers* (1574–1660), 1860: p. 225; see also p. 161. It seems that the promoters of the Darien project expected to raise (? or plant) cochineal (Wafer, 1934: p. lx).

32 Sahagún, 1938: **3**: p. 287.

33 Lee, 1951: p. 210.

34 Guicciardini (1560), 1593: p. 34.

35 Teixeira, 1902: p. 119.

son referred in 1597 to *cotchenella* lying in Pera, and on an earlier occasion (1586) he advised that a consignment should be sent to Cairo.[36] Persia was a recognized market from the beginning of the seventeenth century.[37] The workshops of Ispahan received supplies that had passed through Venice and sometimes Constantinople.[38] The new dyestuff was appreciated particularly in the cities of the eastern provinces (Khorasan and Kirman) where carpet manufacturers had long experience of imported lac.[39] The same cities also acted as entrepôts along routes leading into India. We know that small amounts of cochineal were reaching Sind from Persia before 1618.[40]

The English shipped cochineal to India within a few years of the founding of the factory at Surat in 1612, but demand was limited and the enterprise never flourished. On October 11, 1617, Sir Thomas Roe, writing from Mandoa (Mandor), advised the factors at Surat: "The cochenell mentioned in your invoice, if it will vent at Suratt, it is better passed away than sent up at charge." [41] And again from Lascar (Lashkar) on November 8: "I wish you not to scatter the cochineal; it is no commodity for ordinary markets and it will be wasted in trials . . . it is too dear ever to be a commodity for this country." [42] Another letter in the same year informed the East India Company that "greater quantities of spices will vend in Agra; cochineal is spent, but in mean quantities. . . ." [43] In 1618 cochineal at Ahmadebad was sold below cost and the Company was advised against further shipments.[44] Two years later a consignment had to be sent to Persia "there being no demand . . . at Surat." [45] Not merely was the market small but the Persians retailed the dyestuff at a lower price and effectively served the cities of northern India.[46] Nevertheless the Company persevered in its efforts [47] and demand seems to have risen in the 1660's. Then in 1674–1675 we again hear that little could be sold.[48] The slump was probably partly related to the reforms of the Mogul emperor Aurangzeb who banned the use of *kasumba* (safflowers from which a red dye was prepared) and cochineal in Ahmadabad and Agra and other places. East India Company records of the eighteenth century occasionally refer to cochineal [49] but without implying that it was of any great importance.

At the beginning of the nineteenth century the dyers of Kabul and Herāt (Afghanistan) got supplies of "cochineal" (? *Margarodes polonicus*) from Russia, by way of Bukhārā, and also apparently from Chinese Turkistan (? Yarkand).[50] That available around Herāt itself was "found in the root of a plant that flourishes in a marsh." Later, in Tashkent, "cochineal used for dyeing silk red [was] chiefly brought from Bukhārā, although the insect [was] found in abundance in the spring in the neighbourhood [of Tashkent], on the leaves of the ash, mulberry and other trees." [51] In addition to such local insect dyes, Bukhārā distributed quantities of American cochineal, possibly imported through Canton and traded into Central Asia. Alexander Burnes in 1834 was able to compare cochineal "from the banks of the Oxus" with the American product. He also implied that cochineal "brought from the seaports of India" had largely displaced the Russian dyestuff (supplied by merchants of Bukhārā) in the markets of the Punjab.[52]

Trade between Acapulco and Manila commenced in the 1570's; by 1580 cargoes to the Philippines included Mexican *grana* as well as silver.[53] From the first, some of the dyestuff was probably re-exported. It was available in China *ca.* 1700; the Emperor K'ang-hsi (1662–1722) understood that *ko-tcha-ni-la* was an insect product, produced in the Americas, and obtained from Europeans.[54] It became known as "foreign-red" (*yang hung*).[55] In the 1760's English, French, Dutch,

---

36 Sanderson, 1931: pp. 131, 169.

37 *State Papers* (1617–1621), 1870: pp. 159 (1618), 303 (1619). Newberie ([1581], 1625: p. 1414) referred to *cochonillio* at Shiraz, but this was probably kermes or Polish cochineal or "Armenian red," *Porphyrophora hameli*. Persia had a reputation for exquisite dyeing, especially of silk and carpets. In 1579 Richard Hakluyt (1935: **1**: pp. 137–139) drew up a list of instructions to a London dyer, Morgan Hubblethorne, who was being sent to Persia to obtain information about the methods and the materials employed.

38 *State Papers* (1625–1629), 1884: p. 212 (1626).

39 Edwards, 1953: p. 33. Kermes was more widely employed (Robinson, 1882, under *Sarakhs;* Mumford, 1900: p. 52; Neugebauer and Orendi, 1909: p. 80; Hawley, 1913: pp. 38–39; Pope, 1938–1939: **3**: p. 2460, kermes, lac, and "Armenian red"). John Chardin (1686: p. 351) observed that around Marant (Marand) "they gather *cocheneel* . . . though not in any great quantity . . ." (possibly "Armenian red").

40 Roe (1615–1618), 1899: **2**: p. 488.

41 Foster (ed), 1896–1902: **6**: p. 123. Sir Thomas, the first accredited ambassador to the Mogul court from Great Britain, also referred to cochineal "in the Persian Seray (*caravansarai*)."

42 *Ibid.*, p. 149. See also Roe, 1899: **2**: p. 488 (1618).

43 Foster (ed), 1896–1902: **6**: p. 254.

44 Foster (ed), 1906–1927: **1**: pp. 20 (February 15, 1618), 54 (February 9 and 15, 1619). See also *State Papers* (1617–1621), 1870: p. 258 (March 12, 1619).

45 Foster (ed), 1906–1927: **1**: p. 183 (February 18, 1620).

46 *Ibid.*, p. 169. Persian cochineal is here stated to have come "out of Turkye and Barbarye."

47 *State Papers* (1630–1634), 1892: p. 529 (1634).

48 Fawcett (ed), 1936: pp. 239, 250.

49 Datta (ed), 1958: p. 272 (1747).

50 Elphinstone (1815), 1969: pp. 295–296; Meyendorf (1820), 1826: pp. 245, 250–251; J. G. Gerard, 1833: p. 652; Burnes, 1834: **2**: p. 434; 1842: p. 301; Hayward, 1870: p. 134; Kuropatkin, 1882: p. 74 (Kashgar); Aitchison, 1891: p. 43. Among the Yarkandis, according to Aitchison (1874: p. 213), the word *rang* signified "cochineal dye." Vambéry (1864: p. 428) mentions "materials for red colouring" sent to Persia from Turkistan (Bukhārā) through Herāt. *Cf* Pfister, 1935: p. 46.

51 Schuyler, 1876: **1**: p. 182.

52 Burnes, 1834: **2**: pp. 168–169, 434.

53 Alegre, 1956–1960: **1**: p. 262.

54 K'ang-hsi, 1779: pp. 477–478.

55 Dr. Joseph Needham, *pers. comm.*, 1974.

Swedish, and Danish merchants frequented Canton.[56] Cochineal was then an article of commerce, not only in China, but also in Cambodia, Siam, and Cochin China [57] where presumably it partly displaced lac dye in the manufacture of superior colored silks. Early in the nineteenth century, *grana* dispatched directly from New Spain and South America (? Peru) was still routed through the Philippines; some was sent on to China and some to Bengal and Madras in exchange for piece goods.[58]

The trading activities of the English East India Company extended to China. In 1802 their cochineal "broke the market" and the price in Canton fell by 50 per cent.[59] That imported by private English merchants stood third by value to cotton and woolen goods over the period 1811–1828.[60] However, it was difficult always to match supply and demand, and at various times in the late eighteenth and early nineteenth centuries the product was re-exported from Canton.[61] Items on which duty was payable at Macao in 1833–1834 included cochineal [62] which by this time had found its way to almost every potential market across the width of Asia.

INTRODUCTION OF THE OPUNTIA CACTUS

Opuntia spp. were introduced to Europe some time before 1550, in circumstances unrecorded, but almost certainly as ornamentals or curiosities. Their unusual appearance caused general astonishment [63] and epitomized the strangeness of the new-found lands. Opuntias could be propagated either from seed or from the thick, fleshy joints, as John Gerard (1597) observed.[64] The latter was customary in the New World, so perhaps joints were first carried to Europe. They remain fresh and green for several weeks or even months if carefully stored. Possibly in the sixteenth century, and certainly later, considerable quantities were carried aboard ships and eaten as a vegetable to combat scurvy. Thus apparently the plant was accidentally introduced into many parts of the Old World tropics. At the end of the eighteenth century ships lying off Madras were supplied with joints of opuntia which, by this time, was already well established in India.[65]

Several modern writers have expressed the view that opuntias, whether in the form of seeds or joints, were brought to Spain within a few years of the discovery of the New World.[66] This is possible, indeed likely, for the Spaniards were encouraged to collect new and remarkable species. However, the earliest known descriptions from Europe date from the 1550's, in herbals or, somewhat paradoxically, commentaries on the Classical botanists. The name most commonly applied to the plant was *ficus indica,* that is fig of the Indies. *Ficus* was a term of ancient usage; the "fig of India" of Theophrastus (*ca.* 372–288 B.C.) and Pliny (A.D. 23–79) was apparently the banyan (*F. benghaliensis*).[67] The *higuera de las Indias* of Christoval Acosta (1578) referred to the banana (*Musa* sp.),[68] known from India itself or the East Indies. The *ficus indica* of concern to us here was generally thought to correspond to a plant found, according to the Classical authors, near Opus, capital of Opuntian Locrians, hence the name opuntia. The basis of the false identification seems to lie in the fact that the "plant of Opus" was said to propagate itself, like the prickly pear, by leaves striking root.[69] The name cactus (Gr. *káktos,* L. *cactos*) was applied by the ancients to various thorny plants, particularly the cardoon (*Cynara cardunculus*),[70] and was first adopted as a generic term by Linnaeus.[71]

Francisco López de Gómara (Zaragoza, 1552) observed that "since [the nopal] is now known [in Spain] there is no need to say more about it." [72] The 1558 edition of Matthioli, *Pedacii Dioscoridis Anazarbei de Medica Materia,* published in Venice, contains an illustration of an opuntia (labeled *Opuntia plinij,*

[56] Blair and Robertson (ed), 1903–1909: **48**: pp. 274–275.

[57] *Ibid.,* p. 306. See also Raynal (1770), 1776: **2**: p. 207.

[58] Blair and Robertson (ed.), 1903–1909: **51**: p. 154 (1819–1822); *Anon.,* 1834: p. 92; Mallat, 1846: **2**: p. 295 (1810).

[59] Morse (ed.), 1926–1929: **2**: pp. 388, 390; **3**: pp. 330–331.

[60] Phipps, 1835: pp. 175–183; Hedde and Rondot, 1849: p. 194.

[61] Pritchard, 1957: pp. 131, 133 (calculated from annual figures given in *Report from the Select Committee of the House of Lords appointed to enquire into the present state of the affairs of the East India Company,* Reports from Committees, VI (3), London, 1830: p. 657).

[62] Morrison, 1834: p. 43; Phipps, 1835: p. 165.

[63] Hernández (1571–1576), 1959: **1**: p. 311.

[64] J. Gerard (substantially a translation of Dodonaeus *Stirpium historiae pemptades,* 1583): p. 1329. Cervantes de Salazar ([1554], 1953: p. 61), describing the wonders of the Indies, was one of the first to refer to the way in which the joints strike root.

[65] Wight, 1850: **2**: p. 49; Burkill, 1913: p. 309; 1935: **2**: p. 1596.

[66] Steinheil in Boissier, 1839–1845: **1**: p. 27; Diguet, 1928: p. 121; Alvarez López, 1945: p. 239. Gentry (1972: p. 84) states that the Portuguese "carried *Agave americana* to the Azores in the sixteenth century. From the Azores it was carried to Europe (Linnaeus)." Some opuntias may well have followed the same route.

[67] Theophrastus, 1961–1968: **1**: pp. 53, 313; Pliny, 1961–1968: **2**: p. 519; **4**: p. 17; Dufrené, 1887: p. 33. The description *Ficus Indica* survived into the eighteenth century (Camell [1701], 1704: pp. 49–50; Osbeck, 1771: **1**: p. 171).

[68] C. Acosta, p. 73; also Linschoten, 1598: pp. 98–99. *Cf.* Pliny, 1961–1968: **4**: p. 19.

[69] Theophrastus, 1961–1968: **1**: p. 55; Pliny, 1961–1968: **6**: p. 237 (*circa Opuntem*). See André, 1956: p. 229 and, on the rediscovery of Theophrastus in the early years of the fifteenth century, Schmitt, 1971: pp. 251–270.

[70] Theophrastus, 1961–1968: **2**: pp. 30–31; Pliny, 1961–1968: **6**: p. 233; André, 1956: p. 63.

[71] Linnaeus (1737), 1938: p. 88 ("The word *Opuntia* . . . being rejected . . . I adopted *Cactus,* which is a synonym, is very ancient . . . and not already appropriated"). *Cf.* Linnaeus (1753), 1957: **1**: p. 468.

[72] López de Gómara, 1954: **2**: p. 143; 1964: p. 157.

FIG. 9. *Tuna* (*Opuntia* sp.). Gonzalo Fernández de Oviedo y Valdés, 1535.

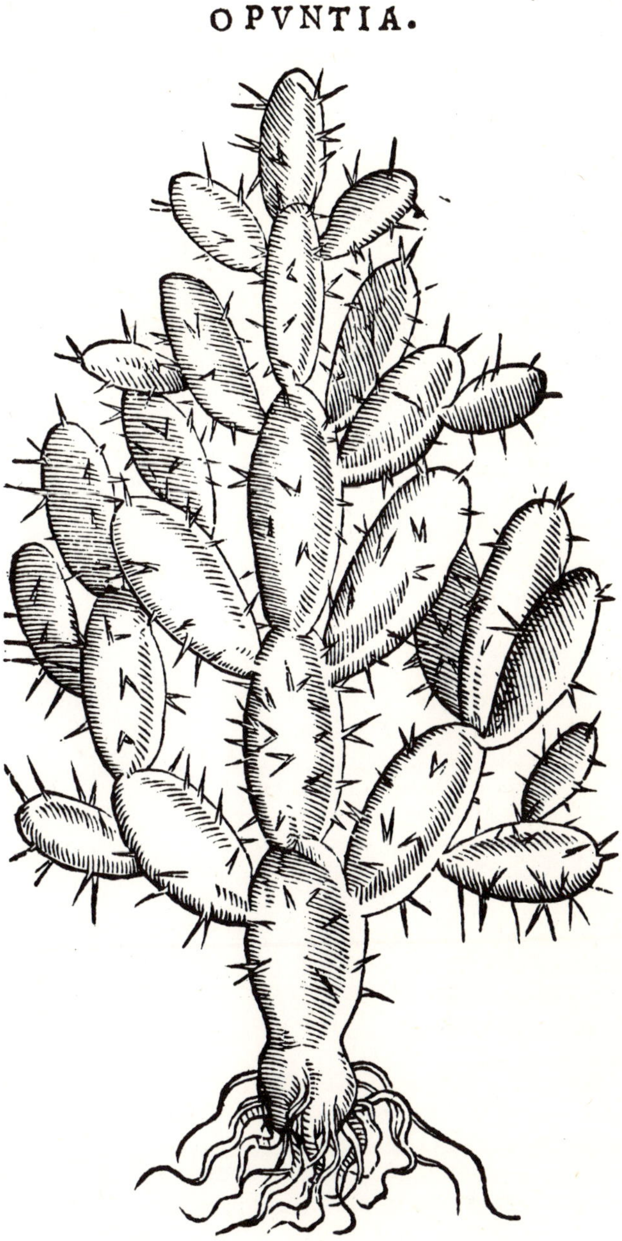

FIG. 10. *Opuntia plinij*. Pietro Andrea Matthioli, 1558.

here, fig. 10) in a section devoted chiefly to the aloe.[73] Nothing earlier based on European experience alone has been discovered.[74] Dodonaeus also illustrated the plant in a book published in 1559.[75] Lobel (1570–1576), Dalechamps (1586) and Camerarius (1588) were among the first of the European herbalists to use the American term *tuna* or *tune,* while still acknowledging the Classical authorities.[76] Francisco Hernández (1571–1576), who studied the flora and fauna of New Spain at first hand, remarked that some authorities believed the identification to be incorrect.[77] The English apothecary and herbalist John Parkinson argued in 1619 that the American species could not be *Opuntia plinij;*[78] nevertheless the error was repeated for a considerable time after this.[79]

Within fifty years or so of their introduction, opuntias were grown in special gardens, some well outside the normal climatic range of the plant. Several gardens in Italy had *Opuntia* spp. by the 1570's.[80] Dodon-

[73] Matthioli, 1558: p. 365 (*Ficus-indica* in the accompanying commentary); the first edition (1544) contains no reference to *Opuntia* or *Ficus-indica. Cf.* Matthioli, 1571: p. 161, 1586: p. 183.

[74] The *Ficus-indica* of Ruellius (1538: p. 207) is not the American plant. Fuchsius (1542) and Tragus (1546, 1552) include nothing of interest. Gesner (1541: p. 41) refers only to "cactus est cinara, suie carduus altilis." The earliest known illustration appears in the Seville (1535) and Salamanca (1547) editions of Oviedo y Valdés *La historia general de las Indias,* lib. VIII, cap. 25 (reproduced here, fig. 9).

[75] Dodonaeus, 1559: p. 443 (*Ficus-indica, Carduus indicus*). There is no reference to *Ficus-indica* in the first (1554*a*) edition of Dodonaeus *Cruÿde Boeck;* but see *ibid.,* 1563: p. 468; 1578: p. 554 ("named by Plinie *Opuntia,* nowe in these dayes *Ficus Indica*"); 1583: pp. 801–802; 1595 [1578]: p. 627.

[76] Lobel, p. 454 (*tuna, tune*); Dalechamps, 2: p. 1795 (*Tune Indorum*); Camerarius, p. 110 (*Tune sive Tunas Indis*). Linocier (1584: p. 292) gives "*Opuntia Plinij,* ou *Ficus Indica.*"

[77] Hernández, 1959: 1: p. 311.

[78] Parkinson, 1628: p. 433. The title page shows a variety of plants, including an opuntia. He states that the description "prickly pear" came into England from Virginia and Bermuda.

[79] Ray, 1688: 2: pp. 1463–1466; Miller, 1731 (under *Opuntia*).

[80] Ramusio [Oviedo y Valdés], 1565: 3: pp. 143–144; Matthioli, 1571: p. 161. *Cf.* Pickering, 1879: p. 659.

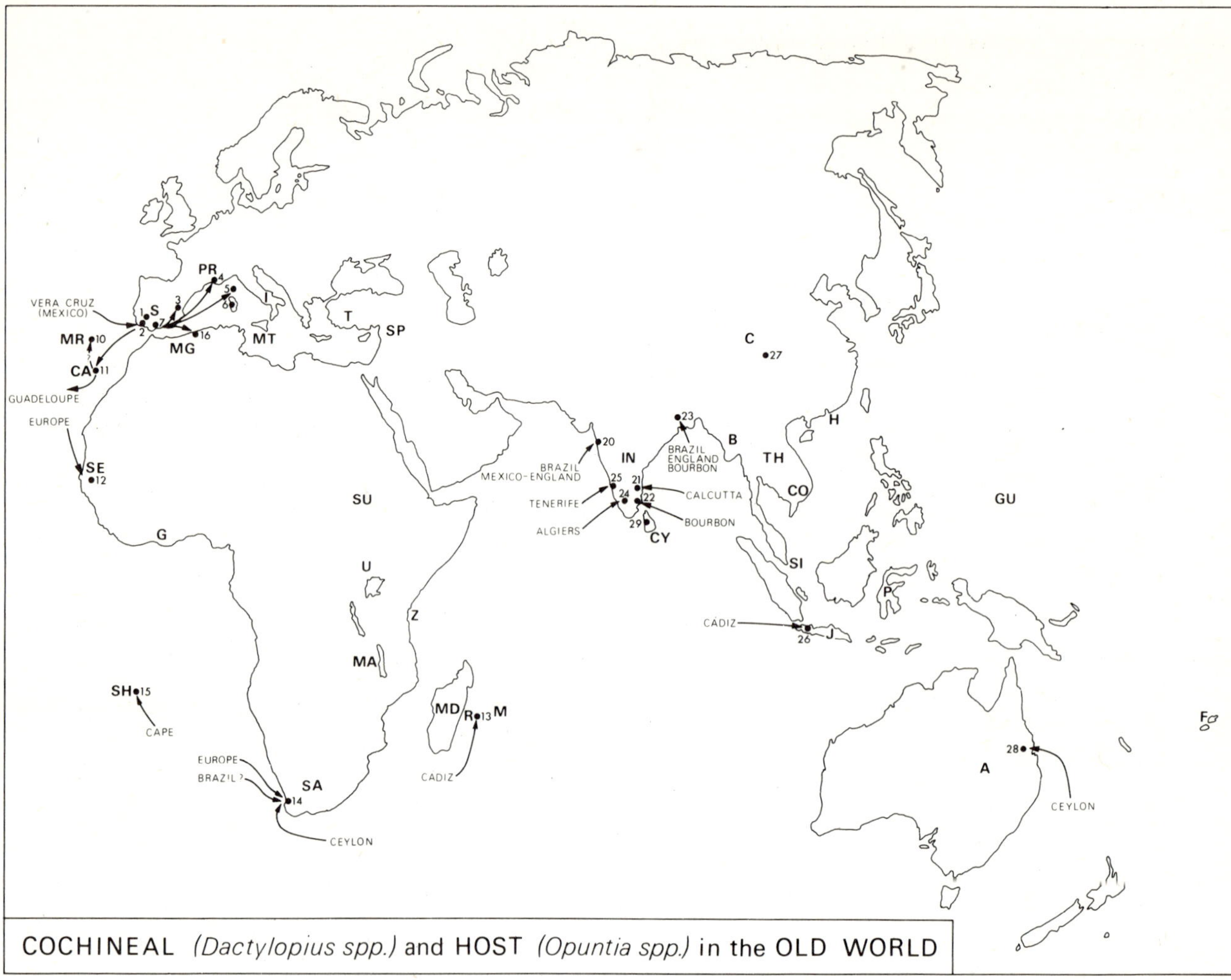

MAP 9.

aeus (1583) mentions Germany and the Low Countries,[81] and the list of specimens in John Gerard's garden at Holborn (1596) includes *"Ficus Indica,"* identified as *O. vulgaris.*[82] In the seventeenth and eighteenth centuries there were gardens with species of *Opuntia* in Holland (Leyden, Hartecamp), France (Paris), England (Eltham, Chelsea, Fulham) and Italy (Vatican, Pisa).[83]

The first opuntias were probably grown in the vicinity of Seville or Cádiz, the terminus of the traffic with the Indies. Subsequent diffusion in southern Spain and other parts of the western Mediterranean may have been relatively rapid. The plant is easily propagated and environmentally undemanding. Spineless varieties provide useful cattle feed,[84] perhaps an important consideration in the stockraising country of Andalusia; and the fruit soon became widely appreciated, especially, it is said, by the Moors.[85]

The popular nomenclature tells us something of the progress of the plant and of the attitude of the local population to what was obviously an exotic. One set of names alludes to the characteristic joint—*higuera de pala,*[86] *figuier de raquette,*[87] and *semelle du Pape* [88] In China, *O. Ficus-indica* is known as "Immortal's hand."[89]

[81] Dodonaeus, 1583: p. 802.

[82] B. D. Jackson (ed.), 1876: pp. 7, 34.

[83] Ray, 1688: 2: pp. 1464, 1466–1467; Tilli, 1723: p. 62; Dillenius, 1732: 2: pp. 395–399; Clifford, 1737: pp. 181–183; Burkill, 1913: p. 288. The municipal garden at Pisa was started in 1544. Burkill (1913: p. 318) identifies the opuntias present in England by the early 18th century.

[84] Grisard, 1896; pp. 480–485.

[85] Shaw, 1738: p. 227; Steinheil in Boissier, 1839–1845: 1: p. 25.

[86] Hernández (1571–1576), 1959: 1: p. 311. Sahagún ([*ca.* 1570], 1963; p. 180) compared the *penca* to a tile.

[87] Sloane, 1725: 2: p. 149; Diderot and d'Alembert, 1751–1757: 3: p. 559. *Raiketra* in Madagascar, according to Grandidier, 1917: p. 201.

[88] Dujardin-Beaumetz and Égasse, 1889: p. 496; Gerth Van Wijk, 1911: p. 919.

[89] Dr. Joseph Needham, *pers. comm.,* 1974. The name has been adopted as the eponym for the whole family of Cactaceae in Chinese taxonomy.

MAP 9. *Dactylopius* spp. and *Opuntia* spp. in the Old World

Reports of the introduction or presence of cochineal insects [numbers have no significance other than identification]

North West Mediterranean province

1 Seville
2 Cádiz
3 Valencia
4 Toulon
5 Corsica
6 Sardinia
7 Malaga

Africa and islands of the Indian and Atlantic Oceans

10 Madeira
11 Tenerife
12 Senegal
13 Bourbon (Réunion)
14 Cape Province
15 St. Helena
16 Algiers

South and East Asia and Australasia

20 Bombay
21 Madras
22 Pondicherry
23 Calcutta
24 Coimbatore
25 Mysore
26 Bantam-Chéribon
27 China
28 Queensland
29 Sri Lanka (Ceylon)

Reports of the introduction or presence of Opuntia spp. (Appendix C)

A Australia
B Burma
C China
CA Canary Islands
CO Cochin China
CY Sri Lanka (Ceylon)
F Fiji
G Ghana
GU Guam
H Hong Kong
I Italy, Sicily
IN India
J Java
M Mauritius
MA Malawi
MD Madagascar
MG Mahgreb
MR Madeira
MT Malta
P Philippines
PR Provence
U Uganda
R Réunion (Bourbon)
S Spain and Portugal
SA South Africa
SE Senegal
SH St. Helena
SP Syria and Palestine
SU Sudan
T Turkey
TH Thailand
Z Zanzibar and Pemba

Such names as *higuera o higuerilla de infierno,*[90] *palmatoria d'infierno,*[91] and *higo de diablo* [92] seem to refer to the bizarre, even supernatural, appearance of the cactus and, or alternatively, to the belief that the (? noncomestible) fruits of the earliest specimens were the cause of illness among the local population.[93] A third set of names associates the opuntia with other cultural groups, thus "fig of the Moors," [94] "of the Christians or Nazarenes," [95] and "of the Jews," [96] all implying strangeness or confusion regarding the origin of the plant.[97] A. Steinheil argued that it was probably taken to Northwest Africa when the Moors were expelled from Spain in 1610 and that the annual pilgrimage to Mecca led to expansion eastward, as far as the Levant and the shores of the Red Sea.

References to opuntias in Asia and Africa south of the Sahara (map 9) are drawn chiefly from floras compiled after 1800. There is, however, good reason to believe that certain species had spread very widely during the two preceding centuries. *O. monacantha* was present in the Cape Province, South Africa, in 1772 and probably much earlier. *O. monacantha* and *O. dillenii* are known to have been well established in India by the 1780's and may have been carried there by the Arabs or the Portuguese in the sixteenth or seventeenth centuries.[98] Perhaps there are undiscovered clues around the sea margins of East Africa and southern Arabia. *Nopalea cochenillifera* was introduced to the sub-continent in 1786 and *O. elatior* and

[90] Cobo (1653), 1956: **1**: p. 200; Ximénez (1722), 1967: p. 267; Burton, 1869: **1**: p. 94 (*Figueira do Inferno*). Montero de Miranda ([*ca.* 1574], 1953: p. 354), describing the province of Verapaz, Guatemala, refers to the *higuera del infierno,* from which an oil (*aceite*) was extracted, but it is not clear that this was a nopal.

[91] Dymock, 1891: **2**: p. 99.

[92] Cobo, 1956: **1**: p. 201 n. 2 (MS. 1757–1761).

[93] Griffiths and Hare (1907: p. 12) mention illnesses attributable to eating *tunas* (in Mexico).

[94] Steinheil in Boissier, 1839–1845: **1**: p. 26; Diguet, 1928: p. 121 (*figuier de Barbarie, higo de Mauro*).

[95] Thomas Shaw (1738: p. 227) gives "*Opuntia* called . . . [*Kermez Nassarah*], The *Fig of the Christians*"; Weber, 1900: p. 5 (*Karmous enn N'sarra*); Low (1924–1934), 1967: **1**: p. 321 (*kermūs en-naṣārā,* Christenfeige); Diguet, 1928: p. 121 (*Karnous enn N'sarra*).

[96] Lobel, 1576: p. 454 (*Judaeorum Tuna Ficifera*).

[97] Also "bastard fig" (Gerth Van Wijk, 1911: p. 919; Steinheil in Boissier, 1839–1845: **1**: p. 26).

[98] Dymock (1891: **2**: p. 99) states that the Portuguese introduced *O. dillenii,* but provides no supporting evidence.

O. nigricans around the turn of the century.[99] Cochineal insects, brought to India in 1795, quickly decimated the host species *monacantha* and *cochenillifera;* on the other hand, *dillenii* and *elatior,* being largely immune to attack and more highly valued as hedge plants (outside rice-growing areas), have flourished and spread up to the present day.

A nopal (? *O. monacantha* or *N. cochenillifera*) is recorded for the Philippines *ca.* 1695.[100] Presumably the plant was brought from New Spain some time after 1565. One of the earliest references to opuntias in China is in a work dating from about 1700.[101] *O. monacantha* and *O. dillenii* are now naturalized along the coast of Kwantung; and in Szech'uan, Hupei, and other interior provinces opuntias are used in the preparation of anthelmintic and insecticidal products.[102] "*Cactus Ficus Indica*" appears in a flora of Indochina published in 1790.[103] The evidence is scattered and somewhat adventitious but it seems reasonable to postulate a process of diffusion from several points of introduction in Asia—the Levant, western India, and the Philippines—during the seventeenth and eighteenth centuries.

SPECULATION CONCERNING THE NATURE OF COCHINEAL

Notwithstanding Europe's long familiarity with kermes and Polish cochineal, and the fact that American cochineal was widely available before the end of the sixteenth century, confusion regarding the nature of the new dyestuff persisted well into the eighteenth century. There were several reasons for this. Accounts of the Indies published (in Spanish) before 1554 were either silent on the matter or to some extent misleading; later and broadly accurate descriptions, also in Spanish, were either ignored or disbelieved; and the Spanish authorities in effect encouraged uninformed speculation by making it very difficult for foreigners to visit the producing areas. Use of the ambiguous term *graine* in English and French further obscured the issue.

Fernández de Oviedo y Valdés, the greatest of the early historians of the Indies, may have started the trail of misunderstanding. His *Historia Natural,* first published in Toledo in 1526, included a description of the opuntia (*tuna*) but no reference to *grana.* The larger *Historia General* (Seville, 1535, and several later editions) mentions only a red pigment prepared from the fruit of a *tuna*-like plant or tree (? *tuna pintadera, tlapalnochtli*) and marketed in cakes.[104] The Franciscan Toribio de Motolinía, writing in the 1530's and '40's, but not published until modern times, appears to have thought that *grana* was a seed.[105] López de Gómara (*Historia General,* published 1552) implied that the dyestuff was obtained from the fruit of the nopal.[106] Cervantes de Salazar came much nearer the truth in a work that appeared in 1554: "Within the flowers [of the *tuna*]," he wrote, "are bred some little worm-like creatures which, when killed in ashes, make the very valuable cochineal, without superior anywhere." [107] The belief that the insect or "worm" fed on the fruit or flower of the opuntia is found in some *Relaciones Geográficas* from Peru (1586, but only recently published) [108] as well as in several later accounts.[109] José de Acosta was probably the first clearly to state in print (1590) that the *gusanillos* were found on the joints of the cactus.[110] He was shortly followed by Vargas Machuca.[111] That *grana* was an insect product was also recognized by Hernández (1571–1576) and Sahagún (*ca.* 1570), each with first-hand knowledge, but whose work was not immediately published.[112] Antonio de Herrera's magisterial *Historia General de los hechos de los Castellanos* appeared in Madrid in 1601–1615 and, one would have supposed, left no room for further doubt—"la cochinilla es cosa

[99] Burkill, 1913: pp. 300, 318.

[100] Mercado, 1936: p. 116 (first published in Blanco, 1877–1883: **4**: p. 39). See Merrill, 1904: p. 24.

[101] *Miscellaneous Notes from South of the Passes* by Wu Chen Fang (Dr. Joseph Needham, *pers. comm.,* 1974).

[102] Dr. Joseph Needham, *pers. comm.,* 1974; F. P. Smith, 1871: p. 291. Wilson (1913: **1**: p. 155) observed that the fruit was not esteemed, but "an extract obtained by boiling the fleshy stems is locally employed as a supposed cure for haemorrhoids." Stuart's *Chinese Materia Medica* (1928: p. 291) gives *Opuntia ficus* (Hsien-jên-chang), adding: "This "fairy palm" is the well known cactus of the plains. It is found in the wilds of Szechuan and Hupeh. It is prescribed, together with licorice . . . and is dried, powdered and mixed with oil. . . ." Opuntias are not mentioned among medicinal plants in the *Pen Ts'ao Kang Mu* (1596) of Li Shih-chên (Read, 1936).

[103] Loureiro, **1**: p. 306. Merrill (1935: p. 276) suggests *Opuntia dillenii* or, somewhat less likely, *O. monacantha.*

[104] Oviedo y Valdés, 1959: **2**: p. 8 ("desta fructa . . . las indias hacen cierta pasta, e córtanla en pedazos cuadrados . . ."). Fester (1940: p. 234) refers to the fruit of *O. soehrensii* as a possible source of red dye in ancient South America; *cf.* F. Herrera, 1934: p. 43 ("sus frutos [de *O. soehrensii*] proportionam una materia tintórea de uso en la culinaria"). Gade (1975: pp. 89, 192) reports that in the upper valley of the Vilcanota "a reddish-purple dye for food and drink is still made from the seeds of *ayrampu* (*O. soehrensii*)."

[105] Motolinía, 1941: p. 197; 1951: p. 258; 1970: p. 86 ("los granos tan grandes como semilla de culantro" [coriander]).

[106] López de Gómara, 1954: **2**: p. 144.

[107] Cervantes de Salazar, 1953: p. 61.

[108] Jiménez de Espada (ed), 1965: **1**: pp. 192 (*gorgojitos*), 234 (*gusanos*).

[109] *Anon.,* 1668: p. 796 (leaves or fruit); Dampier, 1697: pp. 228–229; P. Miller, 1731 (under *Opuntia*).

[110] José de Acosta, 1962: p. 183.

[111] Vargas de Machuca, 1599: p. 165 (*cochinilla*).

[112] Hernández, 1615: p. 31; 1651: pp. 78–79; 1959: p. 315 (*vermiculi, gusanillos,* compared with the coccus [kermes] of the Old World and with a dye insect found in Peru on the roots of *pampínula o teucrio* [? *Teucrium vesicarium*]); Sahagún (first published 1829–1830), 1938: **3**: p. 287 ("unos gusanos que llaman cochinillos"), 1963: p. 239.

FIG. 11. Cochineal, according to Samuel de Champlain (1599–1602), 1936.

viva, a manera de gusano redondo, del tamaño de una lenteja." [113]

Nevertheless, opinion in other parts of Europe in the late sixteenth and the seventeenth centuries was almost equally divided on the question as to whether cochineal was a substance of animal or vegetable origin. Perhaps the earliest, and only sixteenth-century, affirmation of the former view appeared in John Florio's *World of Wordes,* an Italian-English dictionary published in London in 1598. Here *cociniglia* is defined as "a kinde of rich flie or graine coming out of India to dye scarlet with, called Cutchenele." Johanne de Laet (Amsterdam, 1633) described cochineal as "animalculum vel potius insectum," [114] and the English Dominican Thomas Gage, who lived for several years in Mexico and Guatemala, mentioned (1648) the breeding of the "worm" on cultivated "trees." [115] Martin Lister (1671) believed that cochineal was "a sort of kermes"; [116] and John Ray (1688) referred to *vermiculos cochinillae* in discussing the host plant *Ficus-indica.*[117] William Dampier (1697) knew that *grana fina* was an insect product, but was misinformed on matters of breeding and collecting; furthermore, he appears to have thought that the wild cochineal was a kind of seed.[118]

Other writers, from Robert Tomson (1555) to Father Pomet (1694),[119] were persuaded that cochineal was a vegetable substance. The former, who visited New Spain but, unlike Gage, knew little or nothing of the chief areas of production, firmly declared that "Cochinilla is not a worme, or a flye, as some say it is, but a berrie that groweth upon certaine bushes. . . ." [120] Even the usually well-informed Hakluyt (1578) urged that visitors to the Indies should search for "the berrie of Cochenile." [121] In a letter (1609) concerned with the Jamestown (Virginia) settlement, cochineal was associated with "certain unknown kindes of herbes for dieing." [122] Samuel de Champlain's description (1599–1602) owed everything to the imagination; we are informed that "cochineal . . . comes from a fruit the size of a walnut which is full of seed within. It is left to come to maturity until the said seeds are dry, and then it is cut like corn and beaten to have the seed, of which they sow again so as to have more" [123] (fig. 11). Thomas Johnson in his edition of Gerard's *Herbal* (1633) ingeniously combined the essentials of the two conflicting views, observing "upon this plant (*Ficus Indica*) in some parts of the West Indies grow certain excrescences, which in continuance of time turn into insects; and these out-growings are that highly prized *Cochenelle* wherwith they dye colours in graine." [124]

Men of science turned their attention to the problem of the nature of cochineal in the late seventeenth and early eighteenth centuries. This culminated in the work of Anton Van Leeuwenhoek (1704) who, using a microscope, reconstructed the insect from the dried product and correctly observed that cochineal consisted entirely of the females of the species.[125] Even so, the case was not universally accepted. Another Dutchman, Melchior de Ruusscher, took the trouble to obtain supporting documents from Oaxaca and in 1729 published a *Histoire naturelle de la cochenille.* Controversy did not entirely cease until the Abbé Raynal's important account of the East and West Indies appeared in 1770,[126] approximately 250 years after the discovery of cochineal by Europeans.

INTRODUCTION OF THE COCHINEAL INSECT

Proposals to "raise" cochineal formed part of colonial ambitions from the late seventeenth century. The

---

113 A. Herrera, 1934–1956: **9**: p. 154; **5**: p. 492. For other seventeenth-century accounts that state or imply that cochineal was an insect product, see *Memorial* (1620), 1931: p. 498; Dávila Padilla, 1625 (2nd ed.): p. 242; Vázquez de Espinosa (1612–1615), 1942: p. 135; Cobo (1653), 1956: **1**: p. 202; Solis y Rivadeneyra (published 1684), 1753: **1**; p. 247; Burgoa (published 1674), 1934: **1**: p. 279; *Anon.,* 1693: p. 502; Vetancurt (published 1698), 1971: pp. 53–54.

114 Laet, 1633: p. 229. See also Nieremberg, 1635: p. 312.

115 Gage, 1928: p. 167.

116 Lister, 1672: p. 5060.

117 Ray, 1688: **2**: p. 1465.

118 Dampier, 1697: pp. 228–229.

119 Pomet, 1735: **1**: p. 27.

120 Tomson, 1904: p. 358.

121 Hakluyt, 1903–1905: **7**: p. 247 (instructions for Martin Frobisher).

122 Barbour (ed), 1969: **1**: p. 247.

123 Champlain, 1859: p. 25.

124 J. Gerard, 1633: p. 1512 (the first edition, 1597, contains no reference to cochineal). *Cf.* Dillenius, 1732: **2**: p. 399.

125 Leeuwenhoek, 1704: pp. 1614–1618; 1809: pp. 140–147. See also Hire (1694), 1733: pp. 205–206; Plumier (1696), 1733: p. 280; 1704: pp. 211–222; Richter, 1701; Alcazar, 1704: pp. 1765–1769; Ferchault de Réaumur, 1738: **4**: pp. 87–90. Sloane (1707: **1**: p. lxxviii) gives a somewhat garbled (hearsay) account of cochineal-culture in Central America.

126 Raynal, 1776: **2**: p. 204. Dr. Johnson (1755) defined cochineal as "an insect gathered upon the *opuntia,* and dried; from which a beautiful red colour is extracted." A. Anderson (1764: **1**: p. 306; **2**: p. 326) described the insect as a lady bird. This was a common misconception (*Anon.,* 1693: p. 503). Note Fr. *cochenille* (cochineal) and *coccinelle* (lady bird).

British and French had chiefly in mind their possessions in the West Indies.[127] The possibility of introducing live insects to southern Europe was also discussed. Antonio de Ulloa (member of a French expedition to South America, 1735–1744) observed that "Andalusia . . . appears to me extremely well suited for breeding cochineal, both from the nature of the climate and the plantation of fig trees, which there attain so great perfection." [128] Lazzaro Spallanzani, writing in the late eighteenth century, stressed the advantages of introducing the cochineal industry to Sicily and the Eolian islands where the "Indian fig" was already abundant.[129] In making the proposal, he referred to the spread of the silkworm and also claimed that certain French provinces had already benefited by the introduction of cochineal. The Spanish ban on the export of the live insect was, so far as we know, first broken in 1777 when Thiéry de Menonville, after a hazardous journey to Oaxaca, brought *grana fina* to Saint Domingue (Haiti).[130] The enterprise failed, for the Mexican species disappeared after Thiéry's death in 1780. By then, however, wild cochineal (the earlier history of which is unknown) had been found in Haiti. This was harvested for a decade or so and consignments were sent to France in 1787–1788 to be tested at the Gobelin establishment near Paris.[131]

The introduction of wild species of *Dactylopius* to the Old World may have occurred adventitiously through Spanish contacts with the Philippines or Portuguese contacts with the Cape and India, but there is no satisfactory evidence for this. The first documented introduction was from Brazil to India in 1795 and stands to the credit of a British army officer. From the first decade of the nineteenth century, and especially in the 1820's and 1830's, cochineal insects were brought to several other areas: to Spain, France, and probably Italy; and to the colonial territories of Algeria, Senegal and Bourbon, Madeira and the Canary Islands, and Java. These deliberate attempts to promote a seemingly profitable industry coincided with or followed the break-up of the Spanish empire in the New World and the parallel development of possessions in Africa and Asia by the British, French, and Dutch. Nowhere was the effort expended immediately rewarded, and the record of failures or partial failures is longer than that of successes. First, the insects, on joints of cactus, had to survive long sea voyages, and in fact it is doubtful whether the more highly domesticated and least robust species (or varieties) were ever successfully introduced. To provide the appropriate host plant was also a problem in some areas, for not all naturalized opuntias were suitable. Knowledge of the cultivation and preparation of cochineal had then to be disseminated, sometimes in the face of local opposition. Finally, the product was exposed to competition from established areas of production. These difficulties would probably have been insuperable without the active interest of various official and semi-official bodies. Botanical gardens were used to nurture and establish the insects and their hosts. Agricultural societies gave encouragement and support, and numerous pamphlets appeared to advise prospective farmers. Ultimately, the most striking success was achieved in the Canary Islands where physical conditions were particularly appropriate and the economic situation, following the collapse of vine-growing, facilitated rapid growth. In Algeria and Java the industry also made considerable headway, taking advantage of assured markets in France and Holland, but elsewhere it had already declined or even disappeared by the time of the introduction of aniline dyes in the 1860's.

MEDITERRANEAN LANDS

Some obscurity surrounds the first introduction of *Dactylopius* sp. to southern Europe. It is claimed that this occurred in 1806 (four years before the Hidalgo uprising in Mexico), when live insects were brought to Cadiz and also to Toulon.[132] The Toulon experiment, at least, was a total failure. The possibility of "introducing" cochineal to Provence was again discussed in 1856.[133]

A more successful attempt to establish the industry in Europe took place in 1820. Pedro José Carazo, an official of Vera Cruz but born in Malaga, dispatched nopals covered with *grana fina* to the Real Sociedad Económica de Amigos del País in Cádiz.[134] The insects were probably first reared in a garden set aside "for the cultivation of valuable foreign plants" which in 1827 included *Opuntia* sp. complete with cochineal.[135] Meanwhile the Spanish Cortes had expressed approval of the enterprise (1822) and the Society itself published an *Instrucción para el cultivo del nopal y cria de la cochinilla de América* (1825). From Cádiz colonies of *Dactylopius* sp. were sent to agricultural societies in Seville, Malaga, Murcia, and Valencia; thence they were apparently distributed *gratis* to prospective cultivators. Such operations met with some resistance from those who regarded cochineal as a harmful pest on fruit-bearing opuntias. The industry never spread far inland, partly for climatic reasons.[136] The most im-

[127] Savary des Bruslons (1723), 1751–1755: p. 519; Diderot and d'Alembert, 1751–1757: **3**: p. 560.

[128] Ulloa (1748), 1760: **1**: p. 347.

[129] Spallanzani, 1809: pp. 246–247. The fruit was an important item in the diet of the poorer sections of the population.

[130] Thiéry de Menonville, 1787, 1812: pp. 753–876; Edelstein, 1958: pp. 1–8.

[131] *Reports,* 1839: p. 21. Labat ([1697] 1742, **4**: pp. 337–356) refers to cochineal in the West Indies, maintaining that the insect fed on the fruit of the cactus.

[132] R. Blanchard, 1883: p. 77; Figuier, 1868: p. 136.

[133] Berteuil, 1856: **2**: p. 392.

[134] *Reports,* 1839: pp. 37, 66; Burkill, 1913: p. 304.

[135] Mackenzie, 1931: **1**: pp. 211–212.

[136] Vega-Grande, 1865: p. 415.

portant area was probably around Malaga where several influential pamphlets on the rearing of cochineal were published.[137] The first nopalry was established in 1824 in the garden of the convent of Santo Domingo. The enterprise immediately attracted the interest of the Junta de Comercio de Malaga and later benefited from the arrival of Mexican *émigrés* with experience of cochineal production.[138] Output increased from 450 lbs in 1836 to 30,000–40,000 lbs in 1850. However, the Spanish product as a whole was unable to compete successfully, either in respect of quality or price, with imported supplies. The same was also true of cochineal produced after *ca.* 1827 in Corsica, Sardinia, and possibly mainland Italy. The Corsican experiment, described as satisfactory in 1844,[139] was two decades later regarded as unsuccessful.[140]

The introduction of cochineal to Algeria followed directly upon the French occupation. The first attempt, foiled by bad weather and the absence of suitable host plants, was made in 1831, the second, which was eventually successful, in 1833.[141] Malaga, Valencia, and Cádiz have each been suggested as the source of supply;[142] in any event, the insects (and apparently *N. cochenillifera*[143]) came from southern Europe. They were first raised in a private *nopalerie,* then in the Jardins d'Hussein Dey and, after some years of neglect (the project was dismissed as ineffectual in 1839[144]), in the Jardin d'acclimatation d'Alger whose director was largely responsible for naturalizing the species. In 1845 a sample of Algerian cochineal was sent for inspection to the Academy of Sciences in Paris,[145] and two years later seventeen kilograms from the experimental garden were auctioned in Marseilles. Thenceforth, with government encouragement, cultivation gradually expanded. According to L. Figuier, there were fourteen *nopaleries* in 1853 in the province of Algiers alone (fig. 12). France was one of the leading importers of cochineal in the middle of the nineteenth century[146] and production in Algeria continued until the dyestuff was effectively displaced in the 1880's.

[137] For example, Presas, 1825, and Moreno y Galindo, 1837, who describe the varieties of nopal found around Malaga and used in cochineal production.

[138] Sermet, 1973: p. 573.

[139] E. Blanchard, 1844: p. 53.

[140] Figuier, 1868: p. 137. Tristram (1898: p. 319) claimed that *"Coccus cacti"* was "cultivated" in several parts of Palestine, particularly around Nablus, but no confirmation of this has been found.

[141] Loze, 1834; Hardy, 1850; Guérin-Méneville, 1850: p. 4; Figuier, 1868: p. 137.

[142] E. Blanchard, 1844: p. 53; Berteuil, 1856: p. 470; R. Blanchard, 1883: p. 83; Sermet, 1973: p. 573. See also Seurat, 1930: pp. 435, 653.

[143] Bory de Saint-Vincent, 1847: pp. 631–632.

[144] Amyot, 1843: p. 628.

[145] Ministre de la Guerre, 1846: p. 221.

[146] 120,000 kilograms or more annually, 1837–1846; 180,000 kilograms in 1848 (Guérin-Méneville, 1850: p. 2).

THE CANARY ISLANDS AND MADEIRA

Here again there is some disagreement on when and by whom cochineal was first introduced. Instructions "para el uso de los labradores canarios," dated 1825, were published in Ciudad de Laguna (Tenerife) in 1826.[147] Then or in 1827 Sabin Berthelot, director of the Jardín de Aclimatación in Orotava, received a batch of insects from the Economical Society in Cádiz.[148] A year or so later the Spanish government, having lost all control of its possessions in the New World, established a breeding station at Santa Cruz de Tenerife. From here and from the Jardín de Aclimatación colonies were sent to various parts of Tenerife and to other islands in the group, notably, judging from later evidence, Gran Canaria, Fuerteventura, Lanzarote, and Hierro. Berthelot also circulated information to local agricultural societies. Host cacti, including *O. Ficus-indica,* were already naturalized, and climatic conditions on sheltered southfacing slopes, such as around Guimar, proved to be nearly ideal. The insects propagated freely and became widely dispersed, but there was little attention to cultivation or to production of the dyestuff until the 1830's. On the contrary, as in Spain itself, the general view of the population was that cochineal was a parasite from which the opuntias (valued for their fruit) and possibly the grape vine suffered. However the situation changed rapidly when a number of more far-sighted proprietors, with official encouragement, demonstrated that cochineal of high quality could be produced and marketed very successfully.[149] The government established a fund to purchase the product which was also temporarily exempted from tithe and export duties. Exports rose from an initial 8.25 lbs in 1831 to about 1,320 lbs in 1833; by 1840 the figure stood at 66,500 lbs and in 1850 at 782,600 lbs.[150] The continued rapid growth after mid-century was related to the destruction in 1852–1853 of the local vines by the fungoid disease *Uncinula necator* (*Erysiphe tuckeri*). Output reached one million lbs by 1855 and climbed to a peak of over six million lbs in 1875. Cochineal was then by far the most important industry in the islands. But within a decade

[147] Megliorini and Santiago de la Cruz. Sanz Antón (1950: p. 80) maintains that cochineal was first introduced in 1824, O. M. Stone (1887: 1: p. 249) that insects were brought from Mexico in 1825. *Cf.* Laboulbène, 1876: p. 191.

[148] E. Blanchard, 1844: p. 52 (1827); Nobrega, 1848–1849: p. 342 (1826); Honegger, 1879: p. 10 (1826); R. Blanchard, 1883: p. 78 (1827); Morris, 1895: p. 62 (1826); M. Herrera, 1919: p. 164 (1827); Born, 1938: p. 222 (1826). The Orotava garden was established in 1795.

[149] Nobrega (1848–1849: p. 345) claimed that in coastal situations three crops per annum or even "four if a separate plantation be kept for breeding only" were possible.

[150] Berthelot, 1851: p. 6; R. Blanchard, 1883: p. 79. Madoz (1845–1850: **8**: p. 255; **10**: p. 70; **14**: p. 701) gives the amounts exported in 1844 and 1845 from Fuerteventura, Lanzarote and Tenerife. *Cf.* Minutoli, 1854: p. 79 (1831–1852); Fritsch and Rein, 1873: pp. 49–52.

FIG. 12. Harvesting cochineal in Algeria. L. Figuier, 1868.

travelers were reporting a substantial decline, particularly in Tenerife where wine-growing again took first place.[151]

Cochineal was brought to Madeira in the late 1830's.[152] C. J. F. Bunbury noted in 1857 that "the culture of the *Opuntia* . . . has lately been much extended [in Madeira], since hopes have been entertained that the rearing of the cochineal insect may prove a valuable branch of industry. . . ."[153] Nevertheless, the Portuguese authorities do not seem to have pursued the matter with much resolution and the results were disappointing.

In the Canary Islands the host cacti were planted on slopes that, in many cases, would have remained unutilized except as goat pasture. The techniques of cultivation and production were borrowed from the New World, via Spain, except that the plantations were more commonly fertilized with imported guano and the work of collecting cochineal was done largely by women. The creatures were killed by exposure to the sun or by roasting in ovens, "small oval structures of cemented stones, with a small door."[154] Olivia Stone also observed the preparation of black cochineal (*negra*), as well as gray (*madres*) and the usual silver (*plateada*). The first-named variety was obtained by mixing the live insects with "black sand in a bag several feet long, which two men swing backwards and forwards until the juice exudes . . . after which the cochineal is dried in the sun and again shaken with sand to give it brilliancy."[155] By 1887, however, the price offered, barely one-tenth of the reported peak, hardly covered the cost of production and there was considerable distress in those rural areas that had converted almost entirely to production of the dyestuff. Exports steadily diminished but have never entirely ceased.[156] It was from Tenerife that "fine cochineal" was introduced in 1846–1848 to Guadeloupe in the French West Indies,[157] the only known instance of diffusion from the Old to the New World.

151 O. M. Stone, 1887: 1: pp. 269, 305.

152 *Reports*, 1839: p. 51; Nobrega, 1848–1849: p. 342, Vega-Grande, 1865: p. 414.

153 Bunbury, 1857: p. 3.

154 O. M. Stone, 1887: 1: p. 204 (near Orotava).

155 *Ibid.*, p. 269. A surface gloss was one of the desired qualities in cochineal. In the importing centers, black cochineal (of whatever origin) was sometimes mixed with powdered talc to give the preferred silver appearance.

156 Sanz Antón (1950: p. 75) gives figures for 1945–1947.

157 Daver, 1849: pp. 15–48.

SUB-SAHARAN AFRICA

The introduction or presence of cochineal has been reported from Senegal and the Cape region of South Africa. According to A. Sébire, insects were taken to the former territory in 1827 and reared on naturalized *O. Ficus-indica.*[158] Whether there was ever any exportable production is not recorded and appears to be unlikely.

A consignment of live cochineal on rooted cactus (probably *O. monacantha*) was sent to India from the Cape in 1837–1838;[159] and a report published in 1839 stated that a wild species was then spreading in southernmost Africa.[160] Insects taken to India from Brazil in 1795 were routed round the Cape, which suggests the possibility of a comparatively early introduction. The above-mentioned report also referred to cochineal brought from Hamburg *ca.* 1832. Probably there were several attempts to establish the insect in the early nineteenth century but without, so far as is known, any economic advantage.[161]

INDIA

The English East India Company became interested in the possibility of raising cochineal in India soon after 1786. In that year Dr. James Anderson of Madras sent to Sir Joseph Banks "samples of a dye-yielding insect which proved to be a species of *Coccus.*"[162] This and the known presence of opuntias in the peninsula have prompted the suggestion that *Dactylopius* sp. had been introduced some time before 1786, possibly by the Portuguese. In fact there is no evidence that the *coccus* in question was cochineal or even that it fed on a species of cactus; and, as I. H. Burkill points out, if *Dactylopius* sp. had been introduced it would presumably have had the effect (later observed) of destroying one or more of the naturalized opuntias.[163] Probably what was sent to England was the lac insect, also a *coccus.* In any event, the Court of Directors of the Company, guided by Banks, drew the conclusion that the rearing of cochineal was possible and forthwith declared their intention of introducing the "true" insect, it was hoped from Brazil.[164]

The task was partly accomplished in 1795 when Captain Neilson, en route for India, called at Rio de Janeiro and there obtained live insects feeding on joints of opuntia. From the Cape he sent colonies to St. Helena and to Bombay, but the greater part was taken to Calcutta, whence a proportion was dispatched to Anderson in Madras. Before the year was out Anderson published *An account of the introduction of American cochineal-insects into Hindustan* (in fact a series of letters on the matter).[165] The insects that arrived in Calcutta were entrusted to William Roxburgh, superintendent of the botanical garden (established in 1763). The accompanying cacti had died in transit and the insects refused to feed on opuntias that had recently been brought to the garden from Manila, China, and Kew. Approximately nine-tenths perished before it was discovered that the creatures lived and propagated successfully on the well-established *O. monacantha.* Anderson acted promptly, pointing out to the President and Governor in Council of Madras, Lord Nobart, "the opportunity . . . of converting the most waste, barren and dry lands in your possession to great advantage, by encouraging the cultivation of the plant, which is common and indigenous (*sic*), the opuntia of our hedges, for rearing cochineal."[166] From the following year (1796) batches of the insects were distributed by the collectors of revenue in the presidency, who were also instructed to prepare nopalries and to offer to purchase the dried product. As a result, about 4,400 lbs of "Madras cochineal" were exported to England in 1797 and as much as 36,400 lbs in 1798.[167] The prize, however, fell short of expectations, for it was soon realized that this was *grana silvestre,* less than half as effective as *fina* but just as costly to produce. Prices on the London market sometimes allowed little profit and purchases were substantially reduced in 1809 (8,000 lbs) and 1810 (4,000 lbs)—twenty years or so before there was any significant production elsewhere in the Old World. In 1813 Indian cochineal shipped to London was said to be "very small [in quantity], not very abundant in colouring matter, and very inferior to any brought from New Spain."[168]

There were several subsequent attempts to introduce domesticated *Dactylopius* to India but, although official bodies and the East India Company offered large rewards, none, it seems, was successful.[169] In 1821 and again the following year insects, brought from Campeche (Yucatán) and reared in the botanical garden

[158] Sébire, 1899: p. 141.
[159] Burkill, 1913: p. 305.
[160] *Reports,* 1839: p. 10.
[161] Green (1912: p. 84) described living specimens of "cultivated cochineal" (*Coccus cacti*) from the Cape, but offered no suggestion as to the date of introduction. Later (1922: p. 359) he referred to the wild sub-species, *D. confusus-capensis* (on *Opuntia monacantha,* but without injurious effect). According to the same authority, *D. indicus* was taken to South Africa in 1913, apparently in an attempt to eradicate *O. monacantha.*
[162] Watt, 1889: 2: pp. 399–400; 1908: p. 347.
[163] Burkill, 1913: p. 319.
[164] Royle, 1840: p. 59.
[165] See also Ainslie, 1826: p. 79.
[166] J. Anderson, 1795: p. 9.
[167] *Reports,* 1839: pp. 87–89; Royle, 1840: p. 60. Buchanan (1807: 3: pp. 399–400) observed the rearing of cochineal, and the destruction of the "aboriginal" cactus, around Arcot in Madras.
[168] Milburn, 1813: 2: p. 208; similarly *Minutes,* 1831: p. 33.
[169] *Dactylopius indicus* was taken to Ceylon at some unknown date; as a result, *O. monacantha* was replaced by *O. dillenii* (Green, 1922: p. 358). Burkill (1935: 2: p. 1587) refers to the (?) recent introduction to India of a species of cochineal insect able to destroy *O. dillenii* and *O. nigricans.*

in Chelsea, were sent to Bombay.[170] Fifteen years later the Bengal branch of the Agricultural and Horticultural Society of India received a supply from England and from the botanical garden in St. Denis, Bourbon (Réunion).[171] An important object of the Society was "the introduction and propagation, if possible, of the fine-grained cochineal insect."[172] However, all the insects perished. The Bourbon colony was established in 1826, directly from Cádiz, and some time before 1838 an unsuccessful effort was made to extend this to Pondicherry.

E. Balfour refers to the introduction of "fine cochineal" to Mysore from Tenerife,[173] with what result is unknown. The final attempt at naturalization took place as late as 1882–1883 when the government of Madras was responsible for importing insects and cactus from Algeria. A nopalry was established at Coimbatore, but again all the insects died.[174] The experiments with cochineal in India lasted, therefore, almost one hundred years, to about the time when the dyestuff was superseded in most parts of the world. During the nineteenth century some home-produced cochineal was used in native manufactures.[175] Kashmir obtained supplies from central India.[176] Today the insect survives locally, particularly in Assam, but it is of no commercial value.

EAST ASIA AND AUSTRALIA

Coccus cacti has been briefly reported from China.[177] Presumably wild insects arrived accidentally on opuntias, perhaps from Assam or through seaboard contacts with Java, the Philippines[177a] or even Mexico itself. No evidence has been found that they were ever exploited, much less cultivated.[178] Cochineal used in dyeing silk in the eighteenth and nineteenth centuries was exclusively of foreign origin.

The introduction of *Dactylopius* sp. to Java owed everything to the enterprise of the colonial authorities.[179] Insects were smuggled out of Cádiz by an agent of the Dutch government in 1827–1828 and a small proportion survived the long sea passage to Batavia. Suitable opuntias, if not already well established, were probably available in the celebrated botanical garden at Buitenzorg (founded 1817).[179a] That cochineal insects were propagated without repeated introductions is remarkable in view of the somewhat unsatisfactory climatic conditions.

Plantations were developed in the drier, northern half of the island, first in the residencies around Batavia (Bantam, Cheribon, Buitenzorg, and Krawang), and before 1839 in central and eastern Java (Samarang, Djapara, Bezoiki). In 1841 there were about one million cochineal-bearing nopals occupying 220 hectares (*bouws*).[180] According to Jules Itier, whose account dates from *ca.* 1846, the industry in Java was organized in two sectors, private and official, the latter more important.[181] The government established nopalries which were handed over to contractors. The contractors repaid advances of capital over a period of ten years by supplying cochineal at a comparatively low price; whatever remained after the annual quota (one-tenth of the value of the advance) had been met could be sold on the open market. Some 45,000 lbs of the dyestuff—the product of about 700,000 nopals—were marketed through official channels at this time, and one private estate produced 10,000 lbs in what was the eighth year of operations. In 1849 the total harvest exceeded 80,000 lbs from about 360 hectares under cactus.[182] The state Culture System, which promoted the cultivation of many tropical products besides cochineal, lasted from *ca.* 1830 to 1865.

Itier described cochineal estates in the provinces of Bantam and Cheribon, western Java. Nopals took eighteen months to mature, were rested for seven months after each harvest, and replaced after eight

---

170 *Reports,* 1839: p. 29. Cochineal was also reared at Kew for a number of years (Jameson, 1827: p. 195).

171 *Reports,* 1839: p. 50; Burkill, 1913: p. 305. Mr. J. Kingsbury has recently drawn my attention to the following report in the *Gentleman's Magazine* for 1803 (part 2: p. 778): "We learn, *via* Bombay, that attempts are making in the Mauritius to raise cochineal, as the Islands abound with plants on which these insects feed. The birds, who defeated a like attempt some years since, by feeding on these insects, have been nearly exterminated." This information is not incorporated in Map 9 or Appendix C.

172 *Anon.,* 1840: p. 206.

173 Balfour, 1885: **1**: p. 759.

174 Watt, 1889: **2**: p. 40; Burkill, 1913: pp. 306, 322. Green (1912: p. 83) examined dried specimens of a cultivated form of cochineal (*Coccus cacti*) from Sind.

175 Dymock, 1891: **2**: p. 100 (Amritsar, 1848).

176 Moorcroft and Trebeck (1819–1825), 1841: **2**: p. 175 (prized above kermes and lac).

177 Signoret, 1875: p. 349.

177a *Dactylopius coccus* does not, however, appear to have been reported from the Philippines. See E. Robinson, 1917: pp. 1–43.

178 Dr. Joseph Needham, pers. comm., 1974. J. G. Gerard (1833: p. 652) speculated on the raising of "cochineal" in China. In 1834 (*Anon.,* p. 92) the view was expressed that "the climate and situation of China and Japan being similar to Mexico, it is probable that the cultivation of the [host] plant and domestication of the insect would be successful in these countries."

179 Monod de Froideville, 1847: pp. 207–284; Waeij, 1859: pp. 37–40; Deventer, 1865–1866: **2**: pp. 400–402, 569, 734–744; **3**: pp. 124–125; Soest, 1869–1871: **2**: p. 236; **3**: pp. 160–161. The "East Indies" of Humboldt (1811: **3**: p. 64) and Jameson (1827: p. 195) refers to India. The Dutch had earlier introduced the silkworm to the area around Batavia (Raffles, 1817: **1**: p. 53).

179a *Opuntia* sp. is not described in Rumphius's *Herbarium Amboinense* [–1670], 1741–1750.

180 Soest, 1869–1871: **3**: p. 160.

181 Itier, 1848: **2**: pp. 339–341. See also Furnival, 1933: p. 9; 1939: pp. 104, 142, 165.

182 Soest, 1869–1871: **3**: p. 161. Bosch (1858: p. 298) gives 58,684 lbs from the private sector (*partikuliere*) and 44,221 lbs from official sources.

years. The insects, reared under straw canopies,[183] completed their life cycle in the comparatively short time of sixty-five to seventy days. They were killed and dried in the sun (fourteen days) or in ovens (six days) and then sorted by women into four grades. The greater part was exported to China or to Holland. Apparently no use was made of cochineal in native manufactures.[184]

Dactylopius indicus (*ceylonicus*) was brought to Queensland, Australia, in 1903 and again in 1913–1914 for the purpose of controlling the spread of the prickly pear.[185] The insect would, however, attack only *O. monacantha* (*vulgaris*), and other species of *Dactylopius,* including *tomentosus* (*opuntiae*), were later (1920–1934) introduced to destroy the chief pest pears, *O. inermis* and *O. stricta*.[186]

POSTSCRIPT

Insects were exploited for industrial and medico-pharmaceutical purposes in both Eurasia and the Americas long before the Age of Discoveries. In the New World, the most intriguing question concerns the part played by the peoples of the Andean region in the initiation and diffusion of cochineal culture. A center of origin in the Peruvian *sierra* or in the coastal zone, in association with important developments in the textile arts, would not be incompatible with later and more striking progress in the domestication of *Dactylopius* sp. in Middle America. In both regions preparation of the insect dye was probably preceded by the tinctorial use of the fruit of the host cactus. In Southeast Asia an initial interest in the resin of *Lakshadia* sp. seems to have been followed by the discovery of lac dye.

Western and central Asia present the most complex situation with respect to insect dyes. Around the European and Mediterranean margins of the region, Polish cochineal and kermes (*Kermococcus vermilis*) were collected, and in Armenia and Turkistan several other species of insects. The general appellation *kirmiz* (or some cognate word) confuses the issue; and so too does the secrecy that surrounded the dyeing of fine woolens, carpets, and silk. Mixtures of natural dyestuffs were often employed and recipes varied even between workshops in the same city. Intricate patterns of regional and local trade evolved over a long period of time. American cochineal, known in parts of the region from the late sixteenth century, appears to have been readily accepted, as earlier lac dye had been accepted, as an important, if costly, addition to the repertoire of reds.

The domestication of dye insects was a rare phenomenon, best exemplified by the cochineals of the New World. Selection and care of the host cacti and protection and management of the breeding population were the means of improvement, which seems to have progressed further than in the case of the wax insect *axin*. In any event, the interest shown is remarkable in view of the generally poor record of the peoples of Middle America with respect to animal domestication. Perhaps part of the explanation lies in the prior use of *Dactylopius coccus* and *Llaveia axin* in medicinal preparations.

The rearing and preparation of cochineal was at all times a labor-intensive occupation and one involving considerable risks. For the best result (important in what was essentially a luxury product) a high level of skill was also necessary. The most satisfactory unit of production was the cactus garden worked in combination with subsistence agriculture. During the greater part of the colonial period not much if anything was gained by plantation methods. The latter only came to the fore in the changed conditions of the nineteenth century, when cochineal culture was expanded beyond the traditional centers under the influence of governments and local agricultural societies, and at a time of freer trade and rising demand.

Within about one hundred years of the conquest of Mexico, cochineal was employed in many of the textile and leather-working cities of Eurasia and North Africa. It also quickly found its way into the *materia medica* of both Europe and the Orient. No other organic product of the Indies could compare in reputation and monetary value. Spain's monopoly of supply, while doubtless irksome to other countries, had, after about 1550, no significant effect on the distribution of the dyestuff. It was handled by merchants of almost every nationality. Durable and unusually concentrated, cochineal was ideally suited to long-distance commerce serving a dispersed and specialized market. However, the inevitable effect of repeated transactions and of multiple customs charges was to increase the already high price; and this, in turn, prompted thoughts of introducing "the rich flie or graine."

The opuntia was brought to Spain sometime before 1550. It spread around the shores of the Mediterranean through local interest in the fruit and in the plant itself as an effective boundary to fields and house plots. From at least the seventeenth century, the fleshy joints were included among ships' provisions and in

[183] Illustrated by Monod de Froideville, 1847: p. 255.

[184] Raffles (1817: 1: p. 170) mentions "a beautiful and lasting scarlet and blood red obtained from the roots of *wóng-kudu*" (*Morinda citrifolia* [*tinctoria*], member of the Rubiaceae family).

[185] *Anon.,* 1915: p. 323; F. Wilson, 1960: p. 52. Wilson states that "in 1787 Governor Phillip imported into Australia from Argentina a cochineal insect, possibly *D. ceylonicus* (*indicus*), for dye production. . . ." (Arthur Phillip, first governor of New South Wales, only arrived early in 1788.) H. G. Baker (1964: p. 166) substitutes the Canary Islands for Argentina. No supporting evidence has been found. According to Tryon (1910: p. 197), in 1871 cochineal insects were "established" on *O. cochinellifera* in the botanical garden, Brisbane.

[186] Tillyard, 1926: p. 174; F. Wilson, 1960: pp. 52–56.

this way, apparently, the plant was introduced to parts of southern Asia and of Africa beyond the Sahara.

In contrast to the host species, the cochineal insect was carried to the Old World at a late date and only with considerable difficulty. The Spaniards were naturally concerned to preserve their monopoly of the dyestuff. Foreign travelers were virtually excluded from the Indies, particularly the interior provinces, which partly explains the confusion in Europe regarding the nature of the product. In any event, there seems to have been no attempt, and certainly no serious attempt, to export the insect before Thiéry de Menonville's clandestine and successful expedition to Oaxaca in 1777. The first insects brought to the Old World for the purpose of setting up an industry (in India) came from Brazil in 1794–1795; and it was not for another quarter of a century that species were effectively introduced to the lands around the western Mediterranean. Even then, probably little would have been gained had it not been for the enthusiasm of particular individuals and the active encouragement and support of official and semi-official bodies. Insects were naturalized over a number of years in experimental gardens; prospective producers, once opposition or indifference had been overcome, had to be instructed in the cultivation and preparation of cochineal; and finally the product had to be marketed in the face of growing competition from the now independent countries of Mexico, Guatemala, and Peru. *Grana* from the New World, and particularly from Oaxaca, was generally preferred on grounds of quality. The industry in Algeria and Java depended upon assured markets in France and Holland respectively. Only in the Canary Islands in the 1850's and 1860's did the annual output of cochineal far exceed the requirements of the mother country. Indeed, by the early 1870's it was more than double that of the New World as a whole and six times that of Oaxaca. But this striking reversal of position occurred within a few years of the world-wide decline of the industry, following the manufacture of aniline dyes. The demand for cochineal has, however, never entirely ceased. Curiously enough, it is now employed in the pharmaceutical industry as a harmless coloring agent, whereas for many centuries, and perhaps initially, it claimed attention partly on account of its assumed therapeutic properties.

APPENDIX

A. REFERENCES TO THE RAISING OF COCHINEAL OR TO THE PRESENCE OF *GRANA* IN MIDDLE AMERICA (maps 2–7)

Place or region
variant spelling in *italics*
* unlocated
(R) region (Oxaca) includes the Mixteca

PRE-HISPANIC

ACHIUTLA (Oaxaca) *Achiotlán, Achioatlán, Achiyotlán*	Clark, *Codex Mendoza and Matrícula de Tributos,* 1938: **1**: p. 79 (Lee, 1948: p. 452; Dahlgren de Jordan, 1963: p. 13).
ATATLAHUACA (Oaxaca) *Atatlahuacca, Atlacuéchahuaxan*	Lee, 1958: p. 452; Barlow, 1949: p. 117.
CAMOTLÁN (Oaxaca)	Clark, *Codex Mendoza and Matrícula de Tributos,* 1938: **1**: p. 78 (Lees, 1948: p. 452).
CHOLULA (Puebla) *Cholollan*	Orozco y Berra (1880), 1960: 1: p. 280.
COIXTLAHUACA (Oaxaca) *Coayxlahuacan, Coaixlahuacan, Coaixtlauacan, Coayxtlahuacan*	Clark, *Codex Mendoza and Matrícula de Tributos,* 1938: **1**: p. 78 (Orozco y Berra [1880] 1960: 1: p. 280; Lee, 1948: p. 452; Barlow, 1949: p. 117; Molíns Fabrega, 1954: p. 319; Dahlgren de Jordan, 1963: p. 13).
COPALA (Oaxaca) *Capotlán, Tzapotlán*	Clark, *Codex Mendoza and Matrícula de Tributos,* 1938: **1**: p. 79 (Lee, 1948: p. 452; Dahlgren de Jordan, 1963: p. 13).
CUAXOMULCO (Oaxaca) *Coaxmulco, Cuauxomulco, Coaxomalo, Cuaxumulco, Coaxomulco*	Clark, *Codex Mendoza and Matrícula de Tributos,* 1938: **1**: p. 78 (Lee, 1948: p. 452).
CUICATLÁN (Oaxaca) *Cuicalta*	Clark, *Codex Mendoza and Matrícula de Tributos,* 1938: **1**: p. 78 (Lee, 1948: p. 452).
CUILAPAN (Oaxaca) *Coplapan, Coyolan, Coyolapan, Cuylapa, Coyalpan Cuilapa*	Clark, *Codex Mendoza and Matrícula de Tributos,* 1938: **1**: p. 78 (Lorenzana, 1770: p. 174; Lee, 1948: p. 452; Barlow, 1949: p. 123; Molins Fabrega, 1954: p. 319; Dahlgren de Jordan, 1963: p. 13).
ETLA (Oaxaca) *Etlán*	Clark, *Codex Mendoza and Matrícula de Tributos,* 1938: **1**: p. 78 (Lee, 1948: p. 452).
GUAXOLOTITLÁN (Oaxaca) *Huazolotitlán, Quauxilotitlán*	Clark, *Codex Mendoza and Matrícula de Tributos,* 1938: **1**: p. 78 (Lee, 1948: p. 452).
HUAJOLÓTIPAC (Oaxaca) *Quatzontepec, Quautzontepec*	Clark, *Codex Mendoza and Matrícula de Tributos,* 1938: **1**: p. 78 (Lee, 1948: p. 452).
HUEJOTZINGO (Tlaxcala) *Huexotzinco*	Orozco y Berra [1880], 1960: 1: p. 280; Clavijero [*ca.* 1780], 1964: p. 43.
JALTEPEC (Oaxaca) *Xaltepec*	Clark, *Codex Mendoza and Matrícula de Tributos,* 1938: **1**: p. 78 (Lee, 1948: p. 452).
MACUILSÚCHIL (Oaxaca) *Macuilxóchitl, Macuilxochic*	Clark, *Codex Mendoza and Matrícula de Tributos,* 1938: **1**: p. 78 (Lee, 1948: p. 452).
MITLATONGO (Oaxaca) *Mictlán*	Clark, *Codex Mendoza and Matrícula de Tributos,* 1938: **1**: p. 78 (Lee, 1948: p. 452).
NOCHIXTLÁN (Oaxaca) *Nocheztlán*	Clark, *Codex Mendoza and Matrícula de Tributos,* 1938: **1**: p. 78 (Lee, 1948: p. 452).
OAXACA *Guaxaca, Guauxacac, Huaxyacac, Uaxyacac*	Clark, *Codex Mendoza and Matrícula de Tributos,* 1938: **1**: p. 78 (Lorenzana, 1770: fig. 3; Lee, 1948: p. 452).
OCOTLÁN (Oaxaca) *Octlán*	Clark, *Codex Mendoza and Matrícula de Tributos,* 1938: **1**: p. 78 (Lee, 1948: p. 452).
OZOLOTEPEC (Oaxaca) *Ocelotepec, Ocelotepeque*	Barlow, 1949: p. 123.
TAMAZOLA (Oaxaca) *Tamazolla, Tamaçolan*	Clark, *Codex Mendoza and Matrícula de Tributos,* 1938: **1**: p. 78 (Lee, 1948: p. 452).
TAMAZULAPAN (Oaxaca) *Tamaçolapan, Tamazolapan, Tamaçulapa*	Clark, *Codex Mendoza and Matrícula de Tributos,* 1938: **1**: p. 78 (Lee, 1948: p. 452).
TEITIPAC (Oaxaca) *Teticpa, Teticpac, Titicapa*	Clark, *Codex Mendoza and Matrícula de Tributos,* 1938: **1**: p. 78 (Lee, 1948: p. 452).
TEOTITLÁN (Oaxaca) *Teocuitlatlán*	Clark, *Codex Mendoza and Matrícula de Tributos,* 1938: **1**: p. 78 (Lee, 1948: p. 452).
TEPOSCOLULA (Oaxaca) *Tepuzculula, Tepuzcululan*	Clark, *Codex Mendoza and Matrícula de Tributos,* 1938: **1**: p. 78 (Lee, 1948: p. 452).

Tepozcolollan, Tepuzcololan		
TEXUPA (Oaxaca) *Texopan, Texocpan, Tejúpam*		Clark, *Codex Mendoza and Matrícula de Tributos*, 1938: 1: p. 78 (Lee, 1948: p. 452; Barlow, 1949: p. 117).
TLACOLULA (Oaxaca) *Tlalcuechahuaya, Tlalcuechahnayan, Tlacuechauayan, Tlacuechahuayan, Tlacochaguaya*		Clark, *Codex Mendoza and Matrícula de Tributos*, 1938: 1: p. 78 (Lee, 1948: p. 452).
TLAXIACO (Oaxaca) *Tlachquiauco, Tlachquiarico*		Clark, *Codex Mendoza and Matrícula de Tributos*, 1938: 1: p. 79 (Lee, 1948: p. 452; Barlow, 1949: pp. 112–113; Molíns Fabrega, 1954: p. 319; Dahlgren de Jordan, 1963: p. 13).
TUXTEPEC (R) *Tochtepec, Tziauhcochuac*		Alva Ixtlilxochitl [*ca.* 1640], 1892: p. 198.
YANHUITLÁN (Oaxaca) *Yancuitlán, Yanguitlán, Zancuitlán, Yanantitlán*		Clark, *Codex Mendoza and Matrícula de Tributos*, 1938: 1: p. 78 (Lee, 1948: p. 452).

1520–1550

ACATLÁN (Puebla)	*ca.* 1548–1550	Paso y Troncoso (ed.) *Papeles de la Nueva España* 1 [*Suma de Visitas*], 1905 (Lee, 1948: p. 464).
AHUACAPAN (Jalisco) *Aguaçapan*	*ca.* 1548–1550	Paso y Troncoso (ed.) *Papeles de la Nueva España* 1 [*Suma de Visitas*], 1905 (Lee, 1948: p. 464).
AUTLÁN (Jalisco)	*ca.* 1548–1550	Paso y Troncoso (ed.) *Papeles de la Nueva España* 1 [*Suma de Visitas*], 1905 (Lee, 1948: p. 464; Dahlgren de Jordan, 1961: p. 392).
CHAZUMPA (Puebla) *Chiaçompa, Chiatzumba*	*ca.* 1548–1550	Paso y Troncoso (ed.) *Papeles de la Nueva España* 1 [*Suma de Visitas*], 1905 (Lee, 1948: pp. 454, 464).
CHICHICAPAN (Oaxaca)	*ca.* 1548–1550	Paso y Troncoso (ed.) *Papeles de la Nueva España* 1 [*Suma de Visitas*], 1905 (Lee, 1948: p. 464).
CHOLULA (Puebla)	*ca.* 1548–1550	Paso y Troncoso (ed.) *Papeles de la Nueva España* 1 [*Suma de Visitas*], 1905 (Lee, 1948: p. 464; Dahlgren de Jordan, 1961: p. 392).
COCULA (Vera Cruz) *Coculco, Calcahualco*	*ca.* 1548–1550	Paso y Troncoso (ed.) *Papeles de la Nueva España* 1 [*Suma de Visitas*], 1905 (Lee, 1948: pp. 454, 464).
HUAJUAPAN (Oaxaca) *Guaxuapa, Guajuapa*	1536	Scholes, 1936: p. 201.
HUANIQUEO (Michoacán) *Guaniqueo*	*ca.* 1548–1550	Paso y Troncoso (ed.) *Papeles de la Nueva España* 1 [*Suma de Visitas*], 1905 (Lee, 1948: p. 464).
HUAZOLOTITLÁN (Oaxaca) *Guazolotitlán, Guaxolotitla*	*ca.* 1548–1550	Paso y Troncoso (ed.) *Papeles de la Nueva España* 1 [*Suma de Visitas*], 1905 (Lee, 1948: p. 464).
IXTEPEJI (Oaxaca) *Ystepexi*	*ca.* 1548–1550	Paso y Troncoso (ed.) *Papeles de la Nueva España* 1 [*Suma de Visitas*], 1905 (Lee, 1948: p. 464).
IXTLÁN (Oaxaca) *Ystlán*	*ca.* 1548–1550	Paso y Troncoso (ed.) *Papeles de la Nueva España* 1 [*Suma de Visitas*], 1905 (Lee, 1948: p. 464).
MIXTLÁN (Jalisco) *Miztlán*	*ca.* 1548–1550	Paso y Troncoso (ed.) *Papeles de la Nueva España* 1 [*Suma de Visitas*], 1905 (Lee, 1948: p. 464).
NOCHISTLÁN (Oaxaca)	*ca.* 1548–1550	Paso y Troncoso (ed.) *Papeles de la Nueva España* 1 [*Suma de Visitas*], 1905 (Lee, 1948: p. 464).
OZOLOTEPEC (Oaxaca)	*ca.* 1548–1550	Paso y Troncoso (ed.) *Papeles de la Nueva España* 1 [*Suma de Visitas*], 1905 (Lee, 1948: p. 464).
PETLAÇINGO (Puebla) *Petalçingo*	*ca.* 1548–1550	Paso y Troncoso (ed.) *Papeles de la Nueva España* 1 [*Suma de Visitas*], 1905 (Lee, 1948: p. 464).
PUEBLA (R)	1534–1540	Dahlgren de Jordan, 1963: p. 16.
PUEBLA	*ca.* 1548–1550	Paso y Troncoso (ed.) *Papeles de la Nueva España* 1 [*Suma de Visitas*], 1905 (Lee, 1948: p. 464).
QUECHOLAC (Puebla) *Quechula, Cachula*	*ca.* 1548–1550	Paso y Troncoso (ed.) *Papeles de la Nueva España* 1 [*Suma de Visitas*], 1905 (Lee, 1948: p. 464; Dahlgren de Jordan, 1961: p. 392).
TAMAZULAPAN (Oaxaca)	*ca.* 1548–1550	Paso y Troncoso (ed.) *Papeles de la Nueva España* 1 [*Suma de Visitas*], 1905 (Lee, 1948: p. 464).
TECAMACHALCO (Puebla) *Temachalco*	*ca.* 1548–1550	Paso y Troncoso (ed.) *Papeles de la Nueva España* 1 [*Suma de Visitas*], 1905 (Lee, 1948: p. 464; Dahlgren de Jordan, 1961: p. 392).
TECOMATE (Jalisco) *Tecomatlán*	*ca.* 1548–1550	Paso y Troncoso (ed.) *Papeles de la Nueva España* 1 [*Suma de Visitas*], 1905 (Lee, 1948: p. 464).
TEITIPAC (Oaxaca)	*ca.* 1548–1550	Paso y Troncoso (ed.) *Papeles de la Nueva España* 1 [*Suma de Visitas*], 1905 (Lee, 1948: p. 464).
TEPEACA (Puebla)	*ca.* 1548–1550	Paso y Troncoso (ed.) *Papeles de la Nueva España* 1 [*Suma de Visitas*], 1905 (Lee, 1948: p. 464; Dahlgren de Jordan, 1961: p. 392).
TETIQUIPA (Oaxaca) *Tiquipa*	*ca.* 1548–1550	Paso y Troncoso (ed.) *Papeles de la Nueva España* 1 [*Suma de Visitas*], 1905 (Lee, 1948: p. 464).

TLACOCHAGUAYA (Oaxaca) *Tlalcuchaguaya*	*ca.* 1548–1550	Paso y Troncoso (ed.) *Papeles de la Nueva España* 1 [*Suma de Visitas*], 1905 (Lee, 1948: p. 464).
TLAXCALA (R)	1531	Herrera [*ca.* 1600], 1952: **9**: pp. 247–248.
TLAXCALA	*ca.* 1548–1550	Paso y Troncoso (ed.) *Papeles de la Nueva España* 1 [*Suma de Visitas*], 1905 (Lee, 1948: p. 464; Hamnet, 1971*a*: p. 10).
VAPA (Oaxaca) *Oapa, Vepanapa*	*ca.* 1548–1550	Paso y Troncoso (ed.) *Papeles de la Nueva España* 1 [*Suma de Visitas*], 1905 (Lee, 1948: pp. 454, 464).
YANHUITLÁN (Oaxaca)	*ca.* 1548–1550	Paso y Troncoso (ed.) *Papeles de la Nueva España* 1 [*Suma de Visitas*], 1905 (Lee, 1948: p. 464).
ZACATECAS (R) [Nueva Galicia]	1531	Herrera [*ca.* 1600], 1952: **9**: p. 234.

1551–1600

AHUATLÁN (Oaxaca)	*ca.* 1580	Paso y Troncoso (ed.) *Papeles de la Nueva España* 4 [*Relaciones geográficas*], 1905 (Lee, 1948: p. 465).
AMECAMECA (Mexico) *Mecameca*	1580–1582	Philips, 1904: p. 435.
ATATLAHUACA (Oaxaca) *Atlatlauca*	*ca.* 1580	Paso y Troncoso (ed.) *Papeles de la Nueva España* 4 [*Relaciones geográficas*], 1905 (Lee, 1948: p. 465; Dahlgren de Jordan, 1963: p. 13).
AUTLÁN (Jalisco)	1552	Dahlgren de Jordan, 1963: pp. 12, 14.
CALPAN (Tlaxcala)	1562	Scholes and Adams (ed.), 1958: p. 96.
Calpa, Calpulapa	*ca.* 1580	Lee, 1948: p. 465.
	ca. 1600	Gómez de Cervantes, 1944: pp. 57, 164; Herrera [*ca.* 1600], 1952: **9**: p. 155.
CHIAPAS (R)	*ca.* 1600	Dahlgren de Jordan, 1961: p. 394; Roys, 1943: p. 110.
CHILA (Oaxaca)	*ca.* 1580	Paso y Troncoso (ed.) *Papeles de la Nueva España* 4 [*Relaciones geográficas*], 1905 (Lee, 1948: p. 465; Cabrero Fernández, 1962: p. 694).
CHILAPA (Oaxaca)	*ca.* 1580	Paso y Troncoso (ed.) *Papeles de la Nueva España* 4 [*Relaciones geográficas*], 1905 (Lee, 1948: p. 465).
CHOLULA (Puebla)	1562	Scholes and Adams (ed.), 1958: p. 96.
	ca. 1600	Herrera [*ca.* 1600], 1952: **9**: p. 155.
COIXLAHUACA (Oaxaca)	1581	Durán, 1967: **2**: p. 185.
GUATEMALA (R)	*ca.* 1600	Alcedo (D.) [*ca.* 1790], 1883: p. 451; Dahlgren de Jordan, 1961: p. 394; Rubio Sánchez, 1961: p. 18.
HUEJOTZINGO (Tlaxcala)	1562	Scholes and Adams (ed.), 1958: p. 96.
Guaxzingo, Huexocingo,	1582	Chilton, 1904: p. 363.
Huejotzingo, Waxazingo, Guaxocingo	*ca.* 1580	Paso y Troncoso (ed.) *Papeles de la Nueva España* 5 [*Relaciones geográficas*], 1905 (Lee, 1948: p. 465).
	ca. 1600	Gómez de Cervantes, 1944: pp. 57, 165; Herrera, [*ca.* 1600], 1936: **5**: p. 145; 1952: **9**: p. 155.
ICPATEPEQUE (Oaxaca) *Icpatepec*	*ca.* 1580	Paso y Troncoso (ed.) *Papeles de la Nueva España* 4 [*Relaciones geográficas*], 1905 (Lee, 1948: p. 465; Cabrero Fernández, 1962: p. 694).
IXITLÁN (Puebla) *Icxitlán*	*ca.* 1580	Paso y Troncoso (ed.) *Papeles de la Nueva España* 4 [*Relaciones geográficas*], 1905 (Lee, 1948: p. 465).
MACUILSÚCHIL (Oaxaca)	*ca.* 1580	Paso y Troncoso (ed.) *Papeles de la Nueva España* 4 [*Relaciones geográficas*], 1905 (Lee, 1948: p. 465).
MALINALTEPEC (Oaxaca)	*ca.* 1580	Paso y Troncoso (ed.) *Papeles de la Nueva España* 4 [*Relaciones geográficas*], 1905 (Lee, 1948: p. 465; Dahlgren de Jordan, 1963: p. 13).
*MANALA	*ca.* 1600	Herrera [*ca.* 1600], 1952: **9**: p. 155.
MICHOACÁN (R)	*ca.* 1575	Zorita, 1909: p. 275.
	ca. 1600	Herrera [*ca.* 1600], 1934; **1**: p. 72.
NEJAPA (Oaxaca) *Nexapa, Nixapa*	*ca.* 1580	Paso y Troncoso (ed.) *Papeles de la Nueva España* 4 [*Relaciones geográficas*], 1905 (Lee, 1948: p. 465).
NICARAGUA (R)	1575	MacLeod, 1973: p. 172.
	ca. 1600	Alcedo (D.) [*ca.* 1780], 1883: p. 451; Dahlgren de Jordan, 1961: p. 392, 395; Rubio Sánchez, 1961: p. 18.
OAXACA	1570	Chilton, 1904: p. 364.
	1571–1574	López de Velasco, 1894: p. 229.
	ca. 1600	Herrera [*ca.* 1600], 1952: **9**: p. 155.
OTUMBA-ZEMPOALA (Mexico)	16th cent.	Gibson, 1964: p. 354.
OZOLOTEPEC (Oaxaca)	*ca.* 1580	Paso y Troncoso (ed.) *Papeles de la Nueva España* 4 [*Relaciones geográficas*], 1905 (Lee, 1948: p. 465; Dahlgren de Jordan, 1963: p. 13).
PÁTZCUARO (basin) (Michoacán)	16th cent.	Brand, 1951: p. 149.
PIAXTLA (Oaxaca)	*ca.* 1580	Paso y Troncoso (ed.) *Papeles de la Nueva España* 4

Piaztla, Piastla		[*Relaciones geográficas*], 1905 (Lee, 1948: p. 465; Cabrero Fernández, 1962: p. 694).
PUEBLA	1564	Bodenham, 1904: p. 360.
SUCUPO (Yucatán)	1579	Latorre (ed.), 1920: p. 85; Dahlgren de Jordan, 1961: p. 394.
Sucopo, Zusopo		
TABASCO (R)	*ca.* 1600	Alcedo (D.) [*ca.* 1780], 1883: p. 451; Dahlgren de Jordan, 1961: p. 392.
TECAMACHALCO (Tlaxcala)	1568	Chilton, 1904: p. 363.
	ca. 1600	Gómez de Cervantes, 1944: pp. 57, 164; Herrera [*ca.* 1600], 1952: **9**: p. 155.
TEOTITLÁN (Oaxaca)	*ca.* 1580	Paso y Troncoso (ed.), *Papeles de la Nueva España* **4** [*Relaciones geográficas*], 1905 (Lee, 1948: p. 465).
Teutitlán		
TEPEACA (Tlaxcala)	1568	Chilton, 1904: p. 363.
	ca. 1600	Gómez de Cervantes, 1944: pp. 57, 164.
TEPEUXILA (Oaxaca)	*ca.* 1580	Paso y Troncoso (ed.) *Papeles de la Nueva España* **4** [*Relaciones geográficas*], 1905 (Lee, 1948: p. 465).
Tepeucila		
TETELA (Morelos)	*ca.* 1580	Paso y Troncoso (ed.) *Papeles de la Nueva España* **4** [*Relaciones geográficas*], 1905 (Lee, 1948: p. 465).
TEXALUCA (Oaxaca)	*ca.* 1580	Paso y Troncoso (ed.) *Papeles de la Nueva España* **4** [*Relaciones geográficas*], 1905 (Lee, 1948: p. 465).
*TIQUIBALON (Yucatán)	1579	Dahlgren de Jordan, 1961: p. 394.
TLACOLULA (Oaxaca)	*ca.* 1580	Paso y Troncoso (ed.) *Papeles de la Nueva España* **4** [*Relaciones geográficas*], 1905 (Lee, 1948: p. 465).
TLAXCALA (? R)	1562	Scholes and Adams (ed.), 1958: p. 96.
	1572	Hawks, 1904: p. 380.
	ca. 1600	Gómez de Cervantes, 1944: pp. 57, 164; Herrera [*ca.* 1600], 1952: **9**: p. 155.
*TRANGUYZ	*ca.* 1600	Herrera [*ca.* 1600], 1952: **9**: p. 155.
Tranquiz		
YUCATÁN (R)	1566	Landa, 1966: p. 132.
	ca. 1600	Alcedo (D.) [*ca.* 1780], 1883: p. 451; Dahlgren de Jordan, 1961: p. 394.

1601–1700

AHUACUOTZINGO (Oaxaca)	1610	Mota y Escobar [1608–1624], 1945: p. 265.
Ahuacuauhtzinco		
ATLIHUECIAN (Tlaxcala)	1614	Mota y Escobar [1608–1624], 1945: p. 283.
Atluetzan		
CHIAPAS (R)	*ca.* 1612–1620	Vázquez de Espinosa [*ca.* 1628], 1942: p. 205.
	1620	*Memorial*, 1931: p. 500.
CHIAPA [DE CORZA] (Chiapas)	*ca.* 1630	Gage [1648], 1928: pp. 152, 167.
CHOLULA (Puebla)	1610	Mota y Escobar [1608–1624], 1945: p. 248.
	ca. 1612–1620	Vázquez de Espinosa [*ca.* 1628], 1942: p. 136.
	1615	Torquemada, 1723: **1**: p. 282.
	1620	*Memorial*, 1931: p. 498.
COLIMA (R)	*ca.* 1612–1620	Vázquez de Espinosa [*ca.* 1628], 1942: p. 177.
GUATEMALA (R)	*ca.* 1620	Vázquez de Espinosa [*ca.* 1628], 1942: p. 216.
	ca. 1630	Gage [1648], 1928: p. 342.
GUAZACAPÁN (Guatemala)	1617	Macleod, 1973: p. 172.
HUAJUAPAN (Oaxaca)	1610	Mota y Escobar [1608–1624], 1945: p. 258.
HUEJOTZINGO (Tlaxcala)	*ca.* 1612–1620	Vázquez de Espinosa [*ca.* 1628], 1942: p. 136.
ICPATEPEQUE (Oaxaca)	1610	Mota y Escobar [1608–1624], 1945: p. 260.
JALAPA	*ca.* 1630	Gage [1648], 1928: p. 40.
MICHOACÁN (R)	1615	Torquemada, 1723: **1**: p. 336.
MIXTECA (R)	1620	*Memorial*, 1931: p. 498.
	1633	Laet, p. 261.
NEJAPA (Oaxaca)	*ca.* 1630	Gage [1648], 1928: p. 124.
NICARAGUA (R)	*ca.* 1613	Vázquez de Espinosa [*ca.* 1628], 1942: p. 249.
NUEVA GALICIA (R)	1633	Laet, p. 229.
OAXACA (R)	1620	*Memorial*, 1931: p. 498.
	1653	Cobo, 1956: **1**: p. 202.
	1700	Gemelli Carreri, 1955: **2**: p. 170.
OLINALÁ (Oaxaca)	1610	Mota y Escobar [1608–1624], 1945: p. 264.
PUEBLA (R)	*ca.* 1612–1620	Vázquez de Espinosa [*ca.* 1628], 1942: pp. 133, 166.
SANTA MARÍA NATIVITAS (Oaxaca)	1614	Mota y Escobar [1608–1624], 1945: p. 282.
SILACAYOAPAN (Oaxaca)	1610	Mota y Escobar [1608–1624], 1945: p. 261.
Tzilacayoapan		
SUCHITEPEQUES (Guatemala)	1617	Macleod, 1973: p. 172.
	ca. 1630	Gage [1648], 1928: p. 205.
TEQUIXTEPEC (Oaxaca)	1610	Mota y Escobar [1608–1624], 1945: p. 258.

Tezizitepec		
TLACOTEPEC (Oaxaca)	1610	Mota y Escobar [1608–1624], 1945: p. 260.
TLAPALCATEPEC (Oaxaca)	1674	Burgoa, 1934: **2**: p. 289.
Tlapalcaltepec		
Tlapalcatepeque		
TLAPANCINGO (Oaxaca)	1610	Mota y Escobar [1608–1624], 1945: p. 262.
Tlapantzinco		
TLAXCALA (R)	*ca.* 1612–1620	Vázquez de Espinosa [*ca.* 1628], 1942: pp. 135, 137.
	1615	Torquemada, 1723: **1**: p. 276.
	1620	*Memorial*, 1931: p. 498.
	1633	Laet, p. 229.
TLAXCALA	1684	Solis y Rivadeneyra, 1753: **1**: p. 247.
TOPOYANCO (Tlaxcala)	1614	Mota y Escobar [1608–1624], 1945: p. 281.
TOTONICAPÁN (Guatemala)	1617	Macleod, 1973: p. 172.
XALTEPEC (R)	*ca.* 1612–1620	Vázquez de Espinosa [*ca.* 1628], 1942: p. 179.
Xaltepeque		
YUCATAN (R)	1620	*Memorial*, 1931: p. 501 (Scholes and Roys, 1948: p. 305; Dahlgren de Jordan, 1961: p. 394).
ZACATULA (R)	*ca.* 1612–1620	Vázquez de Espinosa [*ca.* 1628], 1942: p. 177.
ZENDALES (R) (Chiapas)	*ca.* 1612–1620	Vázquez de Espinosa [*ca.* 1628], 1942: p. 207.
ZAPOTECAS (R)	*ca.* 1612–1620	Vázquez de Espinosa [*ca.* 1628], 1942: p. 179.
ZUMPANGO (Oaxaca)	*ca.* 1630	Gage [1648], 1928: p. 119.

1701–1800

AMOLTEPEC (Oaxaca)	1777	Dahlgren de Jordan, 1961: p. 396.
Jamiltepec		
AYOCUESCO (Oaxaca)	1789	Dahlgren de Jordan, 1963: App. I (1).
CHEMAX (Yucatán)	18th cent.	Dahlgren de Jordan, 1961: p. 394.
CHIAPAS (R)	1741	Campbell, 1972: p. 135.
	1786–1789	Alcedo, 1967: **1**: p. 291; **4**: p. 296.
	1794	Alzate y Ramírez, 1852; p. 86.
CHILAPA (Oaxaca) *Chilapapam*	1784 1786–1789	Dahlgren de Jordan, 1963: App. I (1).
CHOLULA (Puebla)	1748	Ulloa, **2**: p. 448.
	1786–1789	Alcedo, 1967: **1**: p. 296.
COATLÁN	1786	Dahlgren de Jordan, 1963: App. I (1).
COYOTEPEC (Puebla)	1788	Dahlgren de Jordan, 1963: App. I (1).
CUALAC (Guerrero)	1787–1788	Dahlgren de Jordan, 1963: App. I (1).
Cualaque		
GUADALAJARA (Jalisco)	1787	Thiéry de Menonville, 1812: p. 786.
GUATEMALA (R)	1753	Diderot and d'Alembert, **3**: p. 560.
	1786–1789	Alcedo, 1967: **1**: p. 296.
GUATULCO (Oaxaca)	1786	Dahlgren de Jordan, 1963: App. I (1).
Huatulco		
GUICHICUBI (Oaxaca)	1785	Dahlgren de Jordan, 1963: App. I (1).
HUAJUAPAN (Oaxaca)	1777	Dahlgren de Jordan, 1961: p. 396.
HUAMUXTITLÁN (Oaxaca)	1785	Dahlgren de Jordan, 1963: App. I (1).
IZÚCAR (Puebla)	1788–1789	Dahlgren de Jordan, 1963: App. I (1).
LACHIGOLO (Oaxaca)	1746–1788	Villaseñor y Sánchez, **2**: p. 172.
MIAHUATLÁN (Oaxaca)	1777	Dahlgren de Jordan, 1961: p. 396.
	1784–1789	Dahlgren de Jordan, 1963: App. I (1).
MINAS (Oaxaca)	1746–1748	Villaseñor y Sánchez, **2**: p. 121.
MIXTECA (R)	1787	Clavijero, **1**: p. 74.
NEJAPA (Oaxaca)	1746–1748	Villaseñor y Sánchez, **2**: p. 149.
Nexapa	1760's	Ajofrín, 1964: **2**: p. 117.
	1784–1789	Dahlgren de Jordan, 1963: App. I (1).
	1794	Alzate y Ramírez, 1852: p. 85.
NOCHIXTLÁN (Oaxaca)	1746–1748	Villaseñor y Sánchez, **2**: p. 169.
	1784–1787 1789	Dahlgren de Jordan, 1963: App. I (1).
NOPALUCA (Puebla)	1788	Dahlgren de Jordan, 1963: App. I (1).
NUEVA GALICIA (R)	1748	Ulloa, **2**: p. 448.
OAXACA (R)	1722	Ximénez, 1967: p. 227.
	1746–1748	Villaseñor y Sánchez, **2**: pp. 111–202.
	1753	Diderot and d'Alembert, **3**: p. 560.
	1786–1789	Alcedo, 1967: **1**: p. 296.
	1787	Thiéry de Menonville, 1812: p. 786.
OAXACA	1748	Ulloa, **2**: p. 448.
	1770	Raynal, 1776: **2**: p. 205.
	1781	Landívar, 1948: p. 193.
	1794	Alzate y Ramírez, 1852: p. 85.

POCHUTLA (Oaxaca)	1784–1789	Dahlgren de Jordan, 1963: App. I (1).
PUEBLA (R)	1746–1748	Villaseñor y Sánchez, **1**: pp. 241–382.
PUEBLA	1786	Dahlgren de Jordan, 1963: App. I (1).
QUIECHAPA (Oaxaca)	1777	Taylor, 1972: p. 94.
SAN MARCOS TUXTLA (Chiapas)	1784–1785 1788	Dahlgren de Jordan, 1963: App. I (1).
	1794	Alzate y Ramírez, 1852: p. 86.
SANTA MARÍA ECATEPEC (Oaxaca)	1776	Vasco, 1963.
SANTA MARÍA LACHIXIO (Oaxaca)	1776	Magán, 1963.
TALIXTAC (Oaxaca) *Talistaca*	1746–1748	Villaseñor y Sánchez, **2**: p. 115.
TAMAZULAPAN (Oaxaca)	1785 1787	Dahlgren de Jordan, 1963: App. I (1).
TEHUACAN (Puebla)	1784–1789	Dahlgren de Jordan, 1963: App. I (1).
TEHUANTEPEC (Oaxaca)	1784–1789	Dahlgren de Jordan, 1963: App. I (1).
TEOTITLÁN DEL CAMINO (Oaxaca)	1784–1787 1789	Dahlgren de Jordan, 1963: App. I (1).
TEOTITLÁN DEL VALLE (Oaxaca)	1784–1789	Dahlgren de Jordan, 1963: App. I (1).
TEPOSCOLULA (Oaxaca)	1777	Dahlgren de Jordan, 1961: p. 396.
TLACOCHAGUAYA (Oaxaca)	1746–1748	Villaseñor y Sánchez, **2**: p. 139.
TLACOLULA (Oaxaca)	1746–1748	Villaseñor y Sánchez, **2**: p. 121.
TLAPA (Guerrero)	1746–1748	Villaseñor y Sánchez, **1**: pp. 330–338.
	1785	Dahlgren de Jordan, 1963: App. I (1).
TLAXCALA (R)	1741	Campbell, 1972: p. 128.
	1748	Ulloa, **2**: p. 448.
	1753	Diderot and d'Alembert, **3**: p. 560.
	1786–1789	Alcedo, 1967: **1**: p. 296.
	1787	Thiéry de Menonville, 1812: p. 786.
VERA CRUZ (R)	1787	Thiéry de Menonville, 1812: pp. 782–783.
VILLA ALTA (Oaxaca)	1760's	Ajofrín, 1964: **2**: p. 117.
	1784–1787	Dahlgren de Jordan, 1963: App. I (1).
XAMILTEPEC (Oaxaca) *Jamiltepec*	1784–1785 1788–1789	Dahlgren de Jordan, 1963: App. I (1).
XICAYÁN (Oaxaca) *Jacayán*	1760's	Ajofrín, 1964: **2**: p. 117.
	1786–1787	Dahlgren de Jordan, 1963: App. I (1).
XILOTEPEC (Oaxaca) *Jilotepec*	1784 1788	Dahlgren de Jordan, 1963: App. I (1).
XOCHIMILCO (Mexico)	*ca.* 1800	Gibson, 1964: p. 569 n.
YANHUITLÁN (Oaxaca)	1784–1789	Dahlgren de Jordan, 1963: App. I (1).
ZIMATLÁN (Oaxaca) *Cimatlán*	1786–1787 1789	Dahlgren de Jordan, 1963: App. I (1).
	1794	Alzate y Ramírez, 1852: p. 85.
ZOQUES (Guatemala)	1722	Ximénez, 1967: p. 227 (Scholes and Roys, 1948: p. 39).

1801–

AMATITLÁN (Guatemala)	1835–1847	Dunlop, pp. 123, 127, 129.
	1841	Stephens, 1846: **1**: pp. 277, 294.
	1878–1883	Stoll, 1886: pp. 44–46.
	1965	Osborne, p. 40.
ANTIGUA (Guatemala)	1829	Thompson, p. 249.
	1841	Stephens, 1846: **1**: pp. 277, 294.
	1847	Dunlop, pp. 127, 129.
	1945	O'Neale, p. 29.
CINTALPA (Chiapas) *Sintalapa*	1845	Pineda, p. 72.
CHIQUIMULA (Guatemala)	1847	Dunlop, p. 135.
CUILAPAN (Oaxaca)	1884	Ober, 1887: p. 529.
EJUTLA (Oaxaca)	1891	Velasco, p. 320.
GUADALAJARA (Jalisco)	1803–1804	Humboldt, 1811, **3**: p. 64.
GUATEMALA (R)	1838	Montgomery, 1839: p. 132.
HONDURAS (R)	1948	Stone, p. 211.
IXTLÁN DE JUÁREZ (Oaxaca)	1891	Velasco, p. 291.
JOCOTENANGO (Guatemala)	1945	O'Neale, p. 29.
MIAHUATLÁN (Oaxaca)	1891	Velasco, p. 322.
MIXTECA (R)	1803–1804	Humboldt, 1811: **3**: p. 64.
NICARAGUA (R)	1948	Kirchhoff, p. 220.
NOCHIXTLÁN (Oaxaca)	1891	Velasco, p. 269.
OAXACA (R)	1803–1804	Humboldt, 1811: **2**: p. 241.
	1878	Ratzel, pp. 263–267.
	1936	Parsons, 1966: p. 45.

OCOTLÁN (Oaxaca)	1891	Velasco, p. 313.
	1928	Diguet, p. 535.
	1930	Schmieder, p. 45.
QUEZALTENANGO (Guatemala)	1965	Osborne, p. 40.
RÍO BABISPE-RÍO GUACHINRO (Sonora)	1829	Hardy, p. 455.
SALAMÁ (Guatemala)	1847	Dunlop, p. 135.
	1935	Osborne, p. 53.
SALVADOR (R)	1948	Stone, p. 211.
TAPALAPA (Chiapas)	1845	Pineda, p. 78.
ZACAPA (Guatemala)	1935	Osborne, p. 53.
	1965	Osborne, p. 40.

B. REFERENCES TO THE RAISING OF COCHINEAL OR TO THE PRESENCE OF *GRANA* (*MACNO*) IN SOUTH AMERICA (map 8).

Regions or provinces:

A	Arequipa	1765	*grana, macno*	Bueno, p. 85.
AY	Aymaraes	1768	*grana, macno*	Bueno, p. 98.
CA	El Chaco	20th cent.	cochineal	Métraux, p. 289.
CH	Chillaos or	1766	*grana*	Bueno, p. 66.
	Lulaychillaos	1786–1789	*grana*	Alcedo, **2**: p. 339.
CM	Chumpivilcas	1768	*grana, macno*	Bueno, p. 102.
	or Chumbivilcas	1786–1789	*grana, macno*	Alcedo, **1**: p. 330.
CN	Cuenca	*ca.* 1580	*grana*	Jiménez de la Espada (ed.), **2**: p. 268.
		1786–1789	*cochinilla*	Alcedo, **1**: p. 405.
CO	Condesuyo	1597	*grana, magno*	Ramírez, p. 41.
CT	Cajatambo	1786–1789	*grana, cochinilla*	Alcedo, **1**: p. 263.
CU	Cuzco	1777–1788	*maccnum*	Ruiz, **1**: pp. 83–84.
		20th cent.	cochineal	Fester, 1954: p. 241.
G	Guamanga or	1568	*grana*	Jiménez de la Espada (ed.), **1**: p. 192.
	Huamanga (Ayacucho)	1777–1788		Ruiz, **1**: p. 83.
		1862	*cochinilla*	Paz Soldan, p. 211.
H	Huanuco	1777–1788	*cochinilla, pillcay*	Ruiz, **1**: p. 130.
IC	Ica	1862	*cochinilla*	Paz Soldan, p. 211.
NG	New Granada	? 16th cent / *ca.* 1810 }	*cochinilla*	Caldas y Tenorio, pp. 430–431.
		1741		Campbell, p. 195.
		ca. 1802		Humboldt, **3**: p. 65.
P	Parinacochas	1767	*grana, magno*	Bueno, p. 77.
		1786–1789	*grana, magno*	Alcedo, **3**: p. 134.
		ca. 1800	*grana, magno*	Haenke, p. 259.
PA	Paraguay	*ca.* 1800	cochineal	Helms, p. 176.
R	Rucanas Antamarcas	1586	*magno*	Jiménez de la Espada (ed.), **1**: p. 247.
S	Surinam(e)	1756		*Reports*, 1829: p. 14.
SP	São Paulo	1817–1820	cochineal	Spix and Martius, **2**: p. 19.
T	Trujillo	1862	*cochinilla*	Paz Soldan, p. 211.
TU	Tucumán	1583	*cochinilla*	Jiménez de la Espada (ed.), **1**: p. 391; Latorre (ed.), 1919: p. 144.
		? 16th cent.	*cochinilla*	Caldas y Tenorio, p. 430.
		ca. 1600	*grana, cochinilla*	Lizarraga, p. 187.
		ca. 1620	cochineal	Vázquez de Espinosa, p. 676.
		ca. 1650	*grana, cochinilla*	León Pinelo, **2**: p. 248.
		1777–1788		Ruiz, **1**: p. 83.
		1827	cochineal	Andrews, **1**: p. 184.
V	Venezuela	1824	cochineal	Hall, p. 31.
		1878	cochineal	Spence, **1**: p. 53.

Environs of towns and cities [numbers have no significance other than identification]

1	Loja or Loxa	*ca.* 1580	*grana*	Jiménez de la Espada (ed.), **2**: p. 293.
		? 16th cent / *ca.* 1810 }	*cochinilla*	Caldas y Tenorio, pp. 430, 432.
		1748	*cochinilla*	Ulloa, **2**: pp. 441, 448.
		1761	*grana, cochinilla*	Jaramillo Alvarado [R. Navarro], pp. 180–181.
		1766	*grana silvestre, cochinilla*	Alcedo (D.), p. 45.
		1786–1789	*cochinilla*	Alcedo, **2**: p. 329.
		1858	*cochinilla*	Villavicencio, p. 444.

2	Ambato	? 16th cent *ca.* 1810	*cochinilla*	Caldas y Tenorio, pp. 430, 432.
		1748	*cochinilla*	Ulloa, **2**: pp. 441, 448.
		1761	*grana, cochinilla*	Jaramillo Alvarado [R. Navarro], pp. 180–181.
		1786–1789	*cochinilla*	Alcedo, **1**: p. 50.
3	Cuenca	? 16th cent *ca.* 1810	*cochinilla*	Caldas y Tenorio, pp. 430, 432.
		1761	*grana, cochinilla*	Jaramillo Alvarado [R. Navarro], pp. 180–181.
4	Jaen de Bracamoros	*ca.* 1810	*cochinilla*	Caldas y Tenorio, p. 432.
		ca. 1802	cochineal (wild)	Humboldt, **3**: p. 76.
5	Cajatambo	1764	*grana silvestre*	Bueno, p. 39.
		ca. 1800	*grana silvestre*	Haenke, p. 206.
6	Cúcuta	*ca.* 1810	*cochinilla*	Caldas y Tenorio, p. 432.
7	Pamplona	*ca.* 1810	*cochinilla*	Caldas y Tenorio, p. 432.
8	Tunja	*ca.* 1810	*cochinilla*	Caldas y Tenorio, p. 432.
9	Soacha	*ca.* 1810	*cochinilla*	Caldas y Tenorio, p. 432.
10	Sogomoso	1668–1688	*grana*	Fernández Piedrahita, **1**: p. 15.
11	Conaché	1873	cochineal	Hutchinson, **2**: p. 147.
12	Cajamarca	20th cent.	*grana*	d'Harcourt [1924], 1962: p. 6.
13	Nazca	1586	*magno*	Jiménez de la Espada (ed.), **1**: p. 234.
14	Guananda	1870	cochineal	Orton, p. 108.
15	Tucumán	1748	*cochinilla*	Ulloa, **2**: p. 448.
		1786–1789	*grana*	Alcedo, **4**: p. 118.
16	Talavera	1583	*grana*	Jiménez de la Espada (ed.), **1**: p. 394.
17	Córdoba	1788	*grana*	Quesada (ed.), p. 564.
18	San Luis	1788	*grana*	Quesada (ed.), p. 564.
19	La Plata (Sucre)	1566	*grana, cochinilla*	Jiménez de la Espada (ed.), **2**: p. 74.
		ca. 1628	*grana silvestre*	Vâzquez de Espinosa, p. 655.
20	Valle de Calchaquí	[archaeol.]	*grana, cochinilla*	Ambrosetti, p. 242.
21	Tarma	1777–1788	*cochinilla, pilcay*	Ruiz, **1**: pp. 83, 122.
22	Rio de Janeiro	1792	cochineal	Staunton, **2**: p. 164; Barrow, pp. 81–83.

C. REFERENCES TO THE PRESENCE OF *OPUNTIA* (*NOPALEA, CACTUS*) SPP. IN THE OLD WORLD, EXCLUDING BOTANICAL GARDENS (map 9).

A	Australia	*O. monacantha, stricta, brasiliensis*	Burkill, 1913: p. 292.
		O. monacantha, vulgaris	Bailey, 1913: p. 222.
		O. monacantha, stricta, vulgaris, ficus-indica, tuna, brasiliensis	Ewart, 1930: pp. 781–782.
B	Burma	*O. cochenillifera, monacantha*	Burkill, 1913: pp. 292–293.
C	China	*O. ficus*	Smith, 1871: p. 291.
		O. sp.	Gill, 1880: **2**: pp. 295–296, 322.
		O. sp.	Hosie, 1890: p. 117.
		O. Dillenii	Bretschneider, 1898: p. 242.
		O. Dillenii	Burkill, 1913: pp. 292, 296.
		O. monacantha	J. Needham, *pers. comm.*, 1973.
CA	Canary Islands	*O. Ficus-indica*	Blanchard, 1844: p. 52.
		O. Ficus-indica	Candolle, 1855: **2**: p. 725.
		O. Ficus-indica	Pérez and Sagot, 1865–1866: pp. 136, 233–234.
CO	Cochinchina	*Cactus Ficus Indica*	Loureiro, 1790: **1**: p. 306 (see Merrill, 1935: p. 276).
CY	Ceylon [Sri Lanka]	*O. cochinillifer*	Moon, 1824: p. 38.
		O. Ficus Indica	
F	Fiji	*O. vulgaris*	Parham, 1964: p. 113.
G	Ghana (and coast of West Africa)	*O.* sp.	Hutchinson and Dalziel, 1927–1928: **1**: p. 188.
		O. sp.	Irvine, 1961: pp. 90–91.
GU	Guam	*N. cochenillifera*	Stone, 1970: p. 431.
H	Hong Kong (and New Territories)	*O. Dillenii*	Hance, 1873: p. 104.
		O. Dillenii	Forbes and Hemsley, 1886–1888: p. 323.
		O. Dillenii	Dunn and Tutcher, 1912: p. 114.
I	Italy and Sicily	*O.* sp. ?	Lobel, 1570/1576: p. 454.
		Indica Ficus	Matthioli, 1571: p. 161.
		Ficus Indica	Dodonaeus, 1583: p. 802.
		Tune Indorum	Dalechamps, 1586: **2**: p. 1795.
		Tuna Indis	Camerarius, 1588: p. 110.
		Ficus Indica	Gerard, 1597: p. 1330.

		Ficus Indica	Ray, 1688: **2**: p. 1464.
		O. sp.	Spallanzani [*ca.* 1800], 1809: pp. 246–247.
		O. vulgaris	Steinheil in Boissier, 1839–1845: **1**: p. 26.
		O. Ficus-indica	Candolle, 1855: p. 725.
IN	India	*O.* sp.	Anderson, 1795: p. 9.
		Cactus indicus, chinensis	Roxburgh, 1832: **2**: pp. 475–476.
		O. Dillenii	Wight and Walker-Arnott, 1834: **1**: p. 363.
		O. sp.	Royle, 1840: p. 57.
		O. sp.	Dalzell and Gibson, 1861: p. 39.
		Cactus opuntia	Wight, 1880: **2**: p. 49.
		O. sp.	Balfour, 1885: **1**: p. 757.
		O. sp.	Cooke, 1903: p. 552.
		O. sp.	Burkill, 1913: pp. 289–392.
		O. Ficus-indica	Parker, 1918: p. 257.
		O. sp.	*Reports*, 1839.
J	Java (Netherlands Indies)	*nopal*	Itier, 1848: **2**: pp. 339–342; Daventer, 1865–1866: **2**: pp. 400 ff.; **3**: pp. 124 ff.; Soest, 1869–1871: **2**: p. 236; **3**: pp. 160–161.
		O. cochenillifera, tomentosa, Dillenii, polyantha	Miquel, 1855: **1**: pp. 650–652.
		O. elatior, Dillenii	Burkill, 1913: p. 290.
		N. cochenillifera, O. vulgaris, elatior	Backer and Bakhuizen van den Brink, 1863: **1**: pp. 315–316.
M	Mauritius and Rodriguez	*O. tuna*	Baker, 1877: p. 135.
		O. monacantha	Burkill, 1913: p. 290.
MA	Malawi	*O. vulgaris*	Binns, 1972: p. 60.
MD	Madagascar	*O. monacantha*	Burkill, 1913: p. 290.
MG	North West Africa (Mahgreb)	*Ficus Indica*	Gerard, 1597: p. 1330.
		Opuntia	Shaw, 1738: p. 227.
		O. vulgaris	Steinheil in Boissier, 1839–1845: **1**: p. 25.
		O. sp.	Bory de St. Vincent, 1847: pp. 631–632.
		nopal	Berteuil, 1856: **1**: p. 420.
		O. Ficus-indica	Candolle, 1855: **2**: p. 725; 1882: p. 275.
		O. Ficus-indica	Murbeck, 1905: p. 8.
		O. Ficus-indica	Diguet, 1928: p. 122.
MR	Madeira	*O.* sp.	Bunbury, 1857: p. 3.
MT	Malta	*O. Ficus-indica*	Borg, 1927: p. 171.
P	Philippine Islands	*O. cochenillifera?*	Mercado [*ca.* 1695], 1936: p. 116.
		Cactus opuntia	Blanco, 1837: pp. 414–415.
		O. cochenillifera, Dillenii	Blanco, 1880: **4**: p. 99.
		O. cochenillifera	Merrill, 1903: p. 171; 1912: p. 340.
PR	Provence (southern France)	*O.* sp. ?	Lobel, 1570/1576: p. 454.
		Tune Indorum	Dalechamps, 1586: **2**: p. 1795.
		Tuna Indis	Camerarius, 1588: p. 110.
		Ficus Indica	Ray, 1688: **2**: p. 1464.
		Cactus opuntia	Berteuil, 1856: **2**: p. 393.
U	Uganda	*O.* sp.	Lund and Tallantire, 1962: p. 97.
R	Réunion (Bourbon)	*O. tuna*	Cordemay, 1895: p. 415.
S	Spain and Portugal	*O.* sp. ?	Lobel, 1570/1576: p. 454.
		Ficus Indica	Dodonaeus, 1583: p. 802.
		Tune Indorum	Dalechamps, 1586: **2**: p. 1795.
		Tuna Indis	Camerarius, 1588: p. 110.
		tunal	Acosta [1590], 1962: p. 183.
		Ficus Indica	Gerard, 1597: p. 1330.
		tuna	Ximénez [1722], 1967: p. 267.
		O. sp.	Linnaeus [1753], 1957: p. 468.
		nopal	Alcedo [1786–1789], 1967: **4**: p. 295.
		O. vulgaris	Boissier, 1839–1845; **2**: p. 229.
		O. vulgaris	Steinheil in Boissier, 1839–1845: **1**: p. 26.
		O. Ficus-indica	Candolle, 1855: **2**: p. 725; 1882: p. 275.
SA	South Africa	*O. monacantha, decumana*	Burkill, 1913: p. 290.
		O. sp.	Watt and Breyer-Brandwijk, 1932: pp. 123–124.
SE	Senegal	*O. Ficus-indica*	Sébire, 1899: p. 141.
		O. tuna	Berhaut, 1954: p. 2.
SH	St. Helena	*nopal*	*Reports*, 1839: p. 25.
		O. sp.	Burkill, 1913: p. 295.
SI	Singapore	*O. monacantha*	Burkill, 1935: **2**: p. 1586.

SP	Syria and Palestine	*O. vulgaris*	Steinheil in Boissier, 1839–1845: **1**: p. 26.
		O. Ficus-indica	Candolle, 1855: **2**: p. 725.
		O. sp.	Burkill, 1913: p. 290.
		O. Ficus-indica, vulgaris, cochenillifera	Post, 1932: **1**: pp. 484–485.
SU	Sudan	*O. Dillenii*	Broun and Massey, 1929: p. 106.
T	Turkey	*O. Ficus-indica*	Davis, 1972: p. 209.
TH	Thailand (Siam)	*O. cochenillifera, vulgaris, elatior*	Craig, 1931: p. 782.
		O. monacantha	Burkill, 1935: **2**: p. 1586.
Z	Zanzibar and Pemba	*O. monacantha*	Burkill, 1913: p. 290.
		O. Dillenii	Williams, 1949: p. 385.

BIBLIOGRAPHY OF WORKS CITED

ABEL, C. 1819. *Narrative of a Journey in the interior of China, and of a Voyage to and from that Country in the years 1816 and 1817* (London).

ACOSTA, CHRISTOVAL. 1578. *Tractado de las drogas y medicinas de las Indias Orientales con sus plantas* (Burgos).

ACOSTA, JOAQUÍN. 1848. *Compendio histórico del descubrimiento y colonización de la Nueva Granada en el siglo décimo sexto* (Paris).

ACOSTA, JOSÉ DE. 1880. *The Natural and Moral History of the Indies* [1590] (ed. and trans. C. R. Markham, Hakluyt Society, 2 v. [LX, LXI], London).

—— 1962. *Historia natural y moral de las Indias* [1590] (ed. E. O'Gorman, México).

ADELHOFER, O. (ed.) 1963. *Codex Vindobonensis Mexicanus: I* (Graz).

ADROSKO, R. J. 1968. *Natural Dyes in the United States* (United States National Museum, Bulletin 281, Washington D.C.).

AINSLIE, W. 1826. *Materia Indica* (2 v., London).

AITCHISON, J. E. T. 1874. *Hand-book of the trade products of Leh, with statistics of trade, from 1867 to 1872* (Calcutta).

—— 1891. "Notes to assist in a farther knowledge of the products of western Afghanistan and of North East Persia." *Trans. Botanical Soc. Edinburgh* **18**: pp. 1–228.

AJOFRÍN, FRANCISCO DE. 1958–1959. *Diario del viaje que hizo a la América en el siglo XVIII* (Instituto Cultural Hispano-Mexicano, 2 v. [XII, XIII], México).

ALBERTI, S. 1961. *Diccionari Castellà-Català i Català-Castellà* (Barcelona).

ALCAZAR, (R. P.). 1704. "Observations sur la cochenille et sur un petit limaçon des Indes qui paroit être le murex des anciens." *Mémoires pour l'histoire des sciences et des beaux arts* (Trévoux) **1704**: pp. 1765–1774.

ALCEDO [Y HERRERA], ANTONIO DE. 1815. *Geographical and Historical Dictionary of America and the West Indies* [1786–1789] (trans. G. A. Thompson, 5 v., London).

—— 1967. *Diccionario geográfico historico de las Indias Occidentales ó América* [1786–1789] (ed. C. Pérez-Bustamante, Biblioteca Autores Españoles, 4 v. [CCV–CCVIII], Madrid).

ALCEDO [Y HERRERA], DIONYSIO DE. 1883. *Piraterias y agresiones de los ingleses y de otros pueblos de Europa en la América Española desde el siglo XVI al XVIII deducido de las obras de Alcedo y Herrera, publícalas Justo Zaragoza* [*ca.* 1780] (Madrid).

—— 1915. *Descripción geográfica de la real audiencia de Quito* [1766] (ed. C. A. González Palencia, Hispanic Society of America, Madrid).

ALEGRE, FRANCISCO JAVIER. 1956–1960. *Historia de la provincia de la Compañia de Jesus de Nueva España* [1764–1766] (ed. E. J. Burrus and F. Zubillaga, Bibliotheca Instituti Historici S.J., IX, 4 v., Rome).

ALEJO DE MEAVE, JOAQUÍN. 1831. "Memoria sobre la pintura del pueblo de Olinalan [1791]." *Gacetas de literatura de México* (Puebla) **2**: pp. 213–221.

ALVA IXTLILXOCHITL, FERNANDO DE. 1892. *Obras históricas* [*ca.* 1605] (ed. A. Chavero, 2 v. in 1, México).

ALVARADO, FRANCISCO DE. 1593. *Vocabulario en lengua Mixteca* (México).

ALVAREZ LÓPEZ, E. 1942. "El Dr Francisco Hernández y sus comentarios a Plinio." *Revista de Indias* **3**: pp. 251–290.

—— 1945. "Las plantas de América en la botánica europea del siglo XVI." *Revista de Indias* **6**: pp. 221–288.

ALZATE Y RAMÍREZ, JOSÉ ANTONIO DE. 1794. "Memoria en que trata del insecto grana ò cochinilla [1777]." *Gacetas de literatura de México* **3**: pp. 199–262.

—— 1831. "Memoria en que trata del insecto grana ò cochinilla [1777–1794]." *Gacetas de literatura de México* (Puebla) **3**: pp. 243–318.

—— 1852. "Descripción de la cochinilla Mixteca y de su cria y beneficio [1777–1794]." *Boletín de la Sociedad Mexicana de Geografía y Estadística* **3**: pp. 82–86.

—— 1875. "Cochinilla mixteca [1852]." In: José Maria Pérez Hernández, *Diccionario geográfico, estadística, histórico, biográfico, de industria y commercio de la República Mexicana* (México) **3**: pp. 480–484.

AMBROSETTI, J. B. 1908. *Exploraciones arqueológicas en la ciudad prehistórica de "La Paya."* Publicaciones de la sección antropológica de la facultad de filosofía y letras **3** (Buenos Aires).

AMYOT, C.-J.-B. 1843. *Histoire naturelle des insectes* (Paris).

ANDERSON, ADAM. 1764. *An Historical and Chronological Deduction of the Origins of Commerce from the earliest Accounts to the present time* (2 v., London).

—— 1787–1789. *An Historical and Chronological Deduction of the Origins of Commerce from the Earliest Accounts to the Present Time* (4 v., London).

ANDERSON, JAMES. 1795. *An Account of the Importation of American Cochineal-insects into Hindostan* (Madras).

ANDRÉ, J. 1949. *Étude sur les termes de couleur dans la langue latine,* Études et Commentaires, **7** (Paris).

—— 1956. *Lexique des termes de botanique en Latin.* Études et Commentaires, **23** (Paris).

ANDREWS, JOSEPH. 1827. *Journey from Buenos Ayres through the Provinces of Tucuman and Salta, 1825–6* (2 v., London).

ANON. 1668. "Concerning Cochineel, accompanied with some suggestions for finding out and preparing such like Substances out of other Vegetables." *Philosophical Transactions of the Royal Society of London* **4**, 40: pp. 796–797.

—— 1694. "Observations on the making of Cochineal, according to a relation had from an old Spaniard at Jamaica." *Philosophical Transactions of the Royal Society of London* **17**: pp. 502–503.

—— 1834. *Description of the City of Canton* (Canton).

—— 1840. "Report of the Agricultural and Horticultural Society of India for 1839." *Transactions of the Agricultural and Horticultural Society of India* **7**: pp. 201–213.

—— 1903. *Manuscrit Ramírez: Histoire de l'origine des Indiens qui habitent la Nouvelle Espagne selon leurs traditions* [1586–89]. Recueil de Voyages et de Documents pour servir a l'Histoire de la géographie depuis le XIIIe. jusque à la fin du XVIe. siècle (ed. and trans. D. Charnay, Paris).

—— 1915. "Destruction of Prickly Pear by the Cochineal insect." *Queensland Agricultural Journal* **4**, 4: p. 323.

—— 1917. *Narrative of Some Things of New Spain and of the Great City of Temestitlan, Mexico, written by the Anonymous Conqueror, a Companion of Hernán Cortés* (ed. and trans. M. H. Saville, Cortes Society, I, New York).

—— 1939. *A Chronicle of the Carmelites in Persia and the Papal Mission of the XVIIth and XVIIIth Centuries* (anonymously ed. and trans., 2 v., London).

—— 1956. *Diccionario Castellano-Otomi, Otomi-Castellano.* Cuadernos del Valle del Mezquital, **1**, 1 (Itzmiquilpan).

ARANGO BUENO, T. 1963. *Precolombia* (Bogotá).

ARISTOTLE. 1963. *Minor Works* (ed. and trans. W. S. Hett, Loeb Classics, London).

—— 1965–1970. *Historia Animalium* (ed. and trans. A. L. Peck, Loeb Classics, 2 v., London).

BACKEBERG, C. 1958–1962. *Die Cactaceae: Handbuch der Kakteenkunde* (6 v., Jena).

BACKER, C. A., and R. C. BAKHUIZEN VAN DEN BRINK. 1963–1968. *Flora of Java* (3 v., Groningen).

BAILEY, F. M. 1913. *Comprehensive Catalogue of Queensland plants* (Brisbane).

BAKER, H. G. 1964. *Plants and Civilization* (London).

BAKER, J. G. 1877. *Flora of Mauritius and the Seychelles* (London).

BALFOUR, E. 1885. *The Cyclopaedia of India and of Eastern and Southern Asia* (3 v., 3rd ed., London).

BANCROFT, E. 1813. *Experimental Researches concerning the Philosophy of Permanent Colours* (2 v., London).

BANCROFT, H. H. 1883. *History of the Pacific States of North America* [*Mexico* **3**: 1600–1803] (San Francisco) **6**.

BARBOUR, P. L. (ed.) 1969. *The Jamestown Voyages under the first Charter* (Hakluyt Society, 2nd ser., 2 v. [CXXXVI, CXXXVII], London).

BARLOW, R. H. 1949. *The Extent of the Empire of the Culhua Mexica.* Ibero-Americana, **28** (Berkeley and Los Angeles).

BARNETT, A. 1909. "Étude technologique d'un tissu péruvien antique." *Journal de la Société des Américanistes de Paris,* nouvelle série, **6**: pp. 265–268.

BARRÈRE. P. 1743. *Nouvelle relation de la France équinoxiale* (Paris).

BARRETT, S. A. 1925. *The Cayapa Indians of Ecuador.* Museum of the American Indian, Indian Notes and Monographs, **40** (New York).

BARROW, J. 1806. *A Voyage to Cochinchina in the years 1792 and 1793* (London).

BARTHOLINUS, THOMAS. 1672. *De confectione alchermes* (Hafniae).

BAZANT, J. 1964. "Evolución de la industria textil poblana 1544–1845." *Historia Mexicana* **13**: pp. 473–516.

BEALS, R. L. 1969. "Southern Mexican Highlands and adjacent Coastal Regions." *Handbook of Middle American Indians* (*Ethnology* I), ed. E. Z. Vogt (Austin and London) **7**: pp. 315–328.

BECKMANN, J. 1846. *A History of Inventions and Discoveries* [1783–1805] (ed. W. Francis and J. W. Griffith, trans. A. W. Johnston, 2 v., London).

BENNETT, W. C., and R. M. ZINGG. 1935. *The Tarahumara: An Indian Tribe of Northern Mexico* (Chicago).

BERHAUT, J. 1954. *Flore de Sénegal* (Dakar).

BERLIN, D., D. E. BREEDLOVE, and P. H. RAVEN. 1974. *Principles of Tzeltal Plant Classification: an Introduction to the Botanical Ethnography of a Maya-speaking People of Highland Chiapas* (London and New York).

BERTEUIL, A. 1856. *L'Algérie française* (2 v., Paris).

BERTHELOT, S. 1851. *De l'industrie de la cochenille aux îles Canaries* (Alger).

BERTHELOT, S., and P. B. WEBB. 1836–1850. *Histoire naturelle des Îles Canaries* (3 v. in 7, Paris).

BERTONIO, L. 1612. *Vocabulario de la lengua Aymara* (Juli).

BEYER, H. 1965. *Mito y simbolismo del México antiguo* [1921] (ed. and trans. C. Cook de Leonard, El Mexico Antiguo, X, Sociedad Alemana Mexicanista, México).

BIART, L. 1885. *Les Aztèques. histoire, moeurs, coutumes* (Paris).

BINNS, B. 1972. *Dictionary of Plant Names in Malawi* (Zomba).

BLAIR, E. H., and J. N. ROBERTSON (eds.). 1903–1909. *The Philippine Islands, 1493–1803* (55 v., Cleveland).

BLANCHARD, E. 1844. "Cochenille." *Dictionnaire universel d'histoire naturelle* (Paris) **4**: pp. 49–54.

BLANCHARD, R. 1883. *Les coccidés utiles* (Paris).

BLANCO, M. 1837. *Flora de Filipinas* (Manila).

—— 1877–1883. *Flora de Filipinas* [v. 4 incorporating the work of Ignacio Mercado, Antonio Llanos and others] (4 v. of text, 2 of plates, 3rd ed., Manila).

BOBAN, E. 1891. *Documents pour servir a l'histoire du Mexique* (2 v., Paris).

BODENHAM, ROGER. 1904. *A Voyage made by M. Roger Bodenham to S. John de Ullua in the Bay of Mexico, in the year 1564.* In: Richard Hakluyt, *The Principal Navigations, Voyages, Traffiques and Discoveries of the English Nation* (Glasgow) **9**: pp. 359–360.

BODENHEIMER, F. S. 1928–1929. *Materialen zur Geschichte der Entomologie bis Linné* (2 v., Berlin).

—— 1947. "The Manna of Sinai." *Biblical Archaeologist* **10**: pp. 2–6.

—— 1951. *Insects as Human Food: a Chapter in the Ecology of Man* (The Hague).

—— 1960. *Animal and Man in Bible Lands* (Leiden).

BOISSIER, E. 1839–1845. *Voyage botanique dans le midi de l'Espagne pendant l'année 1837* (2 v., Paris).

BORAH, W. W. 1941. "The Collection of Tithes in the Bishopric of Oaxaca during the 16th Century." *Hispanic American Historical Review* **21**: pp. 386–409.

—— 1943. *Silk Raising in Colonial Mexico.* Ibero-Americana, **20** (Berkeley and Los Angeles).

—— 1949. "Tithe Collection in the Bishopric of Oaxaca, 1601–1867." *Hispanic American Historical Review* **29**: pp. 498–517.

BORAH, W., and S. F. COOK. 1963. *The Aboriginal Population of Central Mexico on the Eve of the Spanish Conquest.* Ibero-Americana, **45** (Berkeley and Los Angeles).

BORG, J. 1927. *Descriptive Flora of the Maltese Islands* (Malta Government, Valetta).

BORHEGYI, S. F. DE. 1959. "Pre-Columbian Cultural Connections between Mesoamerica and Ecuador." *Middle American Research Institute,* Publication 18 (New Orleans), pp. 143–151.

BORN, W. 1938. "Scarlet." *Ciba Review* **1**, **7**: pp. 206–227.

BORY DE SAINT-VINCENT, J. B. G. M. 1826. "Note sur la naturalisation de la cochenille en Espagne." *Annales des sciences naturelles: Zoologie* **8**: pp. 105–107.

—— 1847. "Cochenille." *Encyclopédie moderne* (Paris) **9**: pp. 628–631.

BOSCH, W. 1858. "De vrije en gedwongen arbeid der Javanen *feitelijk* toegelicht." *Tijdschrift voor nederlandsche Indië* **20**, 2: pp. 285–333.

BRAND, D. D. 1951. *Quiroga, a Mexican Municipio* (Smithsonian Institution, Institute of Social Anthropology, Publication 11, Washington D.C.).

—— 1960. *Coalcomán and Motines del Oro; an ex-distrito of Michoacán, México* (The Hague).

—— 1964. "The Status of Ethnozoologic Studies in Meso-America." *XXXV Congreso Internacional de Americanistas* (México, 1962), *Actas y Memorias* (México) **3**: pp. 131–140.

—— 1966. "Cochineal: Aboriginal Dyestuff from Nueva España." *XXXVI Congreso Internacional de Americanistas* (Spain, 1964), *Actas y Memorias* (Sevilla) **2**: pp. 77–91.

BRASSEUR DE BOURBOURG, E. C. 1857–1859. *Histoire des nations civilisées du Mexique et de l'Amerique Centrale* (4 v., Paris).

BRAVO, H. 1937. *Las cactáceas de México* (México).

BRETSCHNEIDER, E. 1871. *On the Knowledge possessed by the Ancient Chinese of the Arabs and Arabian Colonies and other Western Countries* (London).

—— 1898. *History of European Botanical Discoveries in China* (London).

Brinton, D. G. 1890. *Essays of an Americanist* (Philadelphia).

Britton, N. L., and J. N. Rose. 1963. *The Cactaceae* [1919–1922] (4 v. in 2, New York).

Broun, A. F., and R. E. Massey. 1929. *Flora of the Sudan* (London).

Brulley, A. J. 1795. *Discours historique sur la culture du nopal et l'éducation de la cochenille* (Paris).

Bryant, C. 1783. *Flora diaetetica* (London).

Bryce, J. 1877. *Transcaucasia and Ararat* (London).

Buchanan, F. [F. Hamilton]. 1807. *A Journey from Madras through the Countries of Mysore, Canara and Malabar* (3 v., London).

Bueno, C. 1951. *Geografía del Perú virreinal* [*ca.* 1763–1780] (ed. D. Valcarcel, Lima).

Bühler, A. 1948. "Dyeing among primitive peoples." *Ciba Review* **6**, 68: pp. 2478–2512.

Bunbury, C. J. F. 1857. "Remarks on the Botany of Madeira and Tenerife." *Journal of the Proceedings of the Linnean Society of London (Botany)* **1**: pp. 1–35.

Burbidge, N. T. 1963. *Dictionary of Australian Plant Genera* (Sydney).

Burgoa, Francisco de. 1934. *Geográfica Descripción* [1674] (2 v., Publicaciones del Archivo General de la Nacion, **25**, México).

Burkill, I. H. 1913. "Determinations of the Prickly Pears now Wild in India." *Records of the Botanical Survey of India* **4**: pp. 287–322.

—— 1935. *A Dictionary of the Economic Plants of the Malay Peninsula* (2 v., London).

Burland, C. A. (intro.) 1955. *The Selden Roll: An Ancient Mexican Picture Manuscript in the Bodleian Library at Oxford* (Berlin).

—— 1966. *Codex Laud: True Colour Facsimile-Edition of the Old Mexican Manuscript* (Graz).

Burnes, A. 1834. *Travels into Bokhara* (3 v., London).

—— 1842. *Cabool: being a personal narrative of a journey to, and residence in that city, in the years 1836–1838* (London).

Burton, R. F. 1869. *Explorations of the Highlands of Brazil* (2 v., London).

Bustamante, C. M. de. 1821. *Memoria estadística de Oaxaca y descripción del valle del mismo nombre* (Veracruz).

Cabrero Fernández, L. 1962. "La economía basica de los Indios de la región Mixteca à traves de los Relaciones Geográficas de Indias." *XXXIV Internationalen Amerikanistenkongresses* (Wien, 1960), *Akten* (Wien) **1**: pp. 688–694.

Caldas y Tenorio, F. J. de. 1912*a*. "Memoria sobre la importancia del cultivo de la cochinilla [1810]." *Biblioteca de Historia Nacional* **9**: *Obras de Caldas* (Bogotá), pp. 425–447.

—— 1912*b*. "Memoria sobre el modo de cultivar la cochinilla [1810]." *Biblioteca de Historia Nacional* **9**: *Obras de Caldas* (Bogotá), pp. 463–475.

Camell, G. J. 1701. *Descriptiones Fruticum et Arborum Luzonis.* Appendix to John Ray, *Historiae Plantarum* (3 v., 1686–1704, London) **3**.

Camerarius, Joachim [The Younger]. 1588. *Hortus medicus et philosophicus* (Francofurti ad Moenum).

Campbell, John. 1741. *A Concise History of the Spanish America* (London).

Canáls y Martí, J. P. 1768. *Memorias sobre la grana kermes de España que es el coceum, o cochinilla de los antiguos* (Madrid).

—— 1779. *Memorias sobre la púrpura de los antiguos* (Madrid).

Candolle, A. de. 1855. *Géographie botanique raisonnée ou exposition des faits principaux et des lois concernant la distribution géographique des plantes de l'époque actuelle* (2 v., Paris).

—— 1964. *Origin of Cultivated Plants* [1882] (New York).

Cano, Bartholemew. 1904. Letter from Bartholemew Cano to Peter de Tapia, in Sivill, from Mexico, May 5, 1590. In: Richard Hakluyt, *The Principal Navigations, Voyages, Traffiques and Discoveries of the English Nation* (Glasgow) **10**: pp. 166–167.

Carrera, Fernando de la. 1644. *Arte de la lengua de los valles del obispado de Truxillo* (Lima).

Cerna, D. 1932. "The Pharmacology of the Ancient Mexicans." *Annals of Medical History* **4**: pp. 298–305.

Cervantes de Salazar, Francisco. 1953. *Dialogues* [1554] (ed. and trans. C. E. Castañeda, Austin).

Champlain, Samuel de. 1859. *Narrative of a Voyage to the West Indies and Mexico in the years 1599 to 1602* (ed. N. Shaw, Hakluyt Society, XXIII, London).

—— 1922–1936. *The Works of Samuel de Champlain* (ed. H. P. Biggar, Champlain Society Publications, n.s., 6 v., Toronto).

Chardin, John. 1686. *The Travels of Sir John Chardin into Persia and the East Indies* (London).

Charencey, H. de. 1879. "Des couleurs considérées comme symboles des points de l'horizon chez les peuples du Nouveau-Monde." *Actes de la Société Philologique* (Paris) **8**: pp. 5–69.

Chaucer, Geoffrey. 1957. *The Works of Geoffrey Chaucer* (ed. F. N. Robinson, Oxford).

Chaunu, H., and P. Chaunu. 1955–1959. *Séville et l'Atlantique, 1504–1650* (8 v. in 10, Paris).

Chevalier, A. 1937. "Medieval love of Colour." *Ciba Review* **1**, 1: pp. 28–31.

Chevalier, F. 1943. "Les cargaisons des flottes de la Nouvelle Espagne vers 1600." *Revista de Indias* **4**: pp. 323–330.

—— 1963. *Land and Society in Colonial Mexico: the Great Hacienda* (Berkeley and Los Angeles).

Chiao, C. Y., and D. S. Pen. 1940. "China Wax or Insect-wax Industry in Szechuan." *China Journal* **32**, 3: pp. 107–113.

Chilton, John. 1904. *A Notable Discourse of Master John Chilton* [1598]. In: Richard Hakluyt, *The Principal Navigations, Voyages, Traffiques and Discoveries of the English Nation* (Glasgow) **9**: pp. 360–367.

Chipman, D. E. 1967. *Nuño de Guzmán and the Province of Pánuco in New Spain, 1518–1533* (Glendale).

Chopra, R. N., S. L. Nayar and I. C. Chopra. 1956. *Glossary of Indian Medical Plants* (New Delhi).

Cieza de León, Pedro de. 1864. *The Travels of Pedro de Cieza de León, 1532–1540, contained in the First Part of his Chronicle of Peru* (ed. and trans. C. R. Markham, Haklyut Society, XXXIII, London).

Clark, J. Cooper (ed.) 1938. *Codex Mendoza and Matrícula de Tributos* (3 v., London).

Clavijero, F. J. 1787. *The History of Mexico* [*ca.* 1780] (trans. C. Cullen, 2 v., London).

—— 1964. *Historia antigua de México* [*ca.* 1780] (ed. R. P. M. Cuevas, México).

Clifford, George. 1737. *Hortus Cliffortianus: plantas exhibens quas in hortistam vivis quam siccis, Hartecampi in Hollandia* (*Amstelaedami*).

Clusius, Carolus. 1576. *Rariorum aliquot stirpium per Hispanias obseruatarum historiae, libris dvobvs expressa* (Antverpiae).

Cobo, Bernabé. 1956. *Obras* [*Historia del Nuevo Mundo,* 1653] (ed. P. F. Mateos, Biblioteca de Autores Españoles, 2 v. [XCI, XCII], Madrid).

Cockburn, J. 1810. *The Unfortunate Englishman: a Journey Over Land from the Gulph of Honduras to the Great South Sea* [1730] (London).

Cole, W. 1689. *Purpura Anglicanum* (London).

Collett, H. 1902. *Flora Simlensis* (Calcutta).

Columna, Fabius. 1616. *De Purpura* (Romae).

Comstock, J. H. 1916. "Reports on Scale Insects." *Cornell University Agricultural Experiment Station, Bulletin* 372 (Ithaca), pp. 425–603.

Conzemius, E. 1932. *Ethnographical Survey of the Miskito and Sumu Indians of Honduras and Nicaragua* (Smithsonian Institution, Bureau of American Ethnology, Bulletin 106, Washington D.C.).

Cook, O. F. 1925. "Peru as a Centre of Domestication." *Journal of Heredity* **16**, 2: pp. 33–46; 3: pp. 95–110.

Cook de Leonard, C. 1971. "Minor Arts of the Classic Period in Central Mexico." *Handbook of Middle American Indians* (*Archaeology of Northern Mesoamerica,* I), ed. G. F. Ekholm and I. Bernal (Austin and London) **10**: pp. 206–227.

Cooke, T. 1903. *The Flora of the Presidency of Bombay* (2 v., London) **1**.

Cooper, T. T. 1871. *Travels of a Pioneer of Commerce in Pigtail and Petticoats: or, an Overland Journey from China towards India* (London).

Cordemoy, E. J. de. 1895. *Flore de l'Île de la Réunion* (Paris).

Cordry, D., and D. Cordry. 1968. *Mexican Indian Costumes* (Austin and London).

Cortés, Hernán. 1908. *Letters of Cortés* (ed. and trans. F. A. MacNutt, 2 v., New York).

Cotte, J., and Ch. Cotte. 1918. "Le kermès dans l'Antiquité." *Revue archéologique* **7**: pp. 92–112.

Covarrubias, M. 1946. *Mexico South* (New York).

Coxe, W. 1784. *Travels into Poland, Russia, Sweden and Denmark* (3 v., London).

Craig, W. G. 1931. *Florae Siamensis enumeratio* (Bangkok).

Crawford, M. D. C. 1916*a*. "Peruvian Textiles." *American Museum of Natural History, Anthropological Papers* **12**: pp. 107–191.

—— 1916*b*. "Design and Color in Ancient Fabrics." *American Museums Journal* **16**, 7: pp. 417–431.

Curran, C. H. 1937. "Insect Lore of the Aztecs." *Natural History* **39**: pp. 196–203.

Dahlgren de Jordan, B. 1961. "El nocheztli o la grana de cochinilla mexicana." In: *Homenaje a Pablo Martínez del Río* (México), pp. 387–399.

—— 1963. *La grana cochinilla* (ed. B. Dahlgren de Jordan, México).

—— 1966. *La Mixteca: su cultura e historia prehispánicas* (México).

Dalechamps, Jacques, 1586–1587. *Historia generalis plantarum, in libros XVIII* (2 v., Lugduni).

Dalman, G. 1937. *Arbeit und Sitte in Palästina* (7 v., Gütersloh) **5**.

Dalzell, N. A., and A. Gibson. 1861. *The Bombay Flora* (Bombay).

Dampier, William. 1967. *A New Voyage round the World* (London).

—— 1906. *Dampier's Voyages* (ed. J. Masefield, 2 v., London).

Datta, K. K. (ed.) 1958. *Fort William–India House Correspondence.* India Records Series, 1 [1748–1756], (Delhi).

Daver, (M.) 1849. "Mémoire sur la cochenille et son introduction à la Guadeloupe." *Revue Coloniale,* deuxième série, 2, 1: pp. 15–48.

Dávila Padilla, Agustin. 1625. *Historia de la fundación y discurso de la provincia de Santiago de México, de la Orden de Predicadores* (Brusselas).

Davis, P. H. 1972. *Flora of Turkey* (5 v., Edinburgh) **4**.

Debelmas, J. 1975. "Plantes medicinales d'altitude." *Fitoterapia* **3**: pp. 99–110.

Della Vida, G. L. 1941. "On Kirmiz." *Journal of the American Oriental Society* **61**: pp. 287–288.

Deventer, S. Van (ed.) 1865–1866. *Bijdragen tot de kennis van het landelijk stelsel op Java* (3 v., Zalt-Bommel).

Dézallier D'Argenville, A. J. 1742. *Conchyliologie* (Paris).

Díaz del Castillo, Bernal. 1956. *The Discovery and Conquest of Mexico, 1517–1521* [*ca.* 1568] (ed. G. García, trans. A. P. Maudslay, New York).

—— 1963. *Historia verdadera de la conquista de la Nueva España* [*ca.* 1568] (ed. F. Gómez de Orozco *et al.,* México).

Diderot, D., and J. le R. D'Alembert. 1751–1757. *Encyclopédie ou dictionnaire raisonné des sciences, des arts et des metiers* (17 v., Paris).

Diels, L. 1937. *Beiträge zur Kenntnis der Vegetation und Flora von Ecuador,* Bibliotheca Botanica (Stuttgart) **116**: pp. 1–190.

Diguet, L. 1906. "Contribution à l'étude géographique du Mexique précolombien: Le Mixtécapan." *Journal de la Société des Americanistes de Paris,* nouvelle série, **3**: pp. 15–43.

—— 1909. "L'histoire de la cochenille au Mexique." *Journal de la Société des Américanistes de Paris,* nouvelle série, **6**: pp. 75–99.

—— 1928. *Les cactacées utiles du Mexique* (Paris).

Dillenius, Johann Jacob. 1732. *Hortus Elthamensis, seu plantarum rariorum quas in horto suo Elthami in Cantio coluit . . . J. Sherard Gulielmi p.m. Frater, delineationes et descriptiones* (2 v., Londini).

Dingler, M. 1929. "Cochenille und Kermes." *Die Rohstoffe des Tierreichs* **2**: pp. 343–357.

Dioscórides, Pedanius. 1563. [*De Materia Medica*] *Pedacio Dioscórides Anazarbeo* (ed. Andrés de Laguna, Salamanca).

—— 1952–1959. *La Materia Medica de Dioscórides* (ed. C. E. Dubler, 6 v., Barcelona).

Dodonaeus, Rembertus. 1553. *Trium priorum de stirpium historia commentariorum imagines* (Antverpiae).

—— 1554*a*. *Cruÿde Boeck* (Tantwerpen).

—— 1554*b*. *Posteriorum trium R. Dodonaei . . . de stirpium historia commentariorum imagines* (Antwerpiae).

—— 1559. *Commentariorum de stirpium historia imaginum* (2 v., Antwerpiae).

—— 1563. *Cruÿdeboeck* (Thantwerpen).

—— 1578. *A Nievve Herball, or Historie of Plantes* (trans. H. Lyte, Antwerpe).

—— 1583. *Stirpium historiae pemptades sex sive libri XXX* (Antverpiae).

—— 1595. *A New Herball or Historie of Plants* (trans. H. Lyte, London).

Donkin, R. A. 1974. "*Bixa orellana:* 'The Eternal Shrub.'" *Anthropos* **69**: pp. 33–56.

Dorantes de Carranza, Baltasar. 1902. *Sumaria relación de las cosas de Nueva España* [1604] (ed. L. González Obregón, México).

Dronke, P. 1972. "Tradition and Innovation in Medieval Western Colour-imagery." *Eranos* **41**: pp. 51–107.

Dufrené, H. 1887. *La flore Sanscrite* (Paris).

Dujardin-Beaumetz (M)., and E. Égasse. 1889. *Les plantes medicinales, indigènes et exotiques* (Paris).

Dunlop, R. G. 1847. *Travels in Central America* (London).

Dunn, H. 1829. *Guatimala, 1827–1828* (London).

Dunn, S. T., and W. J. Tutcher. 1912. *Flora of Kwangtung and Hong Kong.* Kew Bulletin, Miscellaneous Information, additional series (London) **10**.

Durán, Diego. 1967. *Historia de las Indias de Nueva España e Islas de la Tierra Firme* [1581] (ed. A. M. Garibay, 2 v., México).

Dusenberry, W. H. 1947. "Woollen Manufacture in Sixteenth-century New Spain." *The Americas* **4**: pp. 223–234.

Dymock, W. 1890–1893. *Pharmaeographia Indica* (3 v., London).

Ealand, C. A. 1914. *Insects and Man* (London).

Edelstein, S. M. 1958. "Spanish Red—Thiery de Menonville's *Voyage à Guaxaca.*" *American Dyestuff Reporter* **47**, 1: pp. 1–8.

EDMONSON, M. S. (ed.). 1971. *The Book of Counsel: The Popol Vuh of the Quiche Maya of Guatemala* [*ca.* 1550–1555]. Middle American Research Institute, Publication 35 (New Orleans).

EDWARDS, A. C. 1953. *The Persian Carpet* (London).

EDWARDS, E. 1966. *Painted Walls of Mexico* (Austin).

EICHSTADIUS, LAURENTIUS. 1634. *De confectione alchermes* . . . (Stetini).

ELLIS, JOHN. 1763. "An Account [to Peter Wych] of the Male and Female Cochineal Insects that breed on the Cactus Opuntia, or Indian Fig, in South Carolina and Georgia." *Philosophical Transactions of the Royal Society of London* [for 1762] **52**, 2: pp. 661–666.

ELPHINSTONE, M. 1969. *An Account of the Kingdom of Caubul* [1815] (Graz).

EMMART, E. W. (ed.) 1940. *The Badianus Manuscript* (*Codex Barberini, Latin 241,* Vatican library): *An Aztec Herbal of 1552, by Martin de la Cruz, trans. Juannes Badianus into Latin from Aztec* (Baltimore).

ESTETE, MIGUEL DE. 1924. *Relación que del descubrimiento y conquista del Perú* [1534]. Colección de libros y documentos referentes a la historia del Perú, segunda serie (Lima) **8**: pp. 3–56.

ESTRADA, JUAN DE, and FERNANDO DE NIEBLA. 1955. *Descripción de la provincia de Zapotitlán y Suchitepéquez: año de 1579.* Anales de la Sociedad de Geografía e Historia (Guatemala) **28**: pp. 68–84.

EVANS, C., and MEGGERS, B. J. 1966. "Mesoamerica and Ecuador." *Handbook of Middle American Indians* (*Archaeological frontiers and external connections*), ed. G. F. Ekholm and G. R. Willey (Austin and London) **4**: pp. 243–264.

EWAN, J. 1970. "Plant Collectors in America: backgrounds for Linnaeus." In: *Essays in Biohistory, presented to Frans Verdoorn on the occasion of his 60th birthday* (ed. P. Smit and R. J. Ch. ter Laage, Utrecht), pp. 19–54.

EWART, A. J. 1930. *Flora of Victoria* (Melbourne).

EXELL, A. W. 1944. *Catalogue of the Vascular Plants of S. Tomé* [*with Principe and Annobon*] (London).

FABER, G. A. 1938. "Greek and Roman Superstitions connected with the Colour Red." *Ciba Review* **1**, 9: p. 311.

FAIRLIE, S. 1965. "Dyestuffs in the Eighteenth Century." *Economic History Review,* second series, **17**: pp. 488–510.

FAWCETT, C. H. (ed.) 1936. *The English Factories in India,* n.s., (4 v., Oxford) **1** [1670–1677].

FERCHAULT DE RÉAUMUR, R. A. 1734–1742. *Mémoires pour servir à l'histoire des insectes* (6 v., Paris).

FERNALD, M. E. 1903. *Catalogue of the Coccidae of the World* (Amherst).

FERNÁNDEZ PIEDRAHITA, LUCAS. 1942. *Historia general de las conquistas del Nuevo Reino de Granada* [1668–1688] (ed. G. Otero Muñoz, 4 v., Bogotá).

FESTER, G. A. 1940–1943. "Los colorantes del antiguo Perú." *Archeion* (Roma) **22**: pp. 229–241; **25**: pp. 195–196.

—— 1953. "Einige Farbstoffe Sudamerikanischer Kulturvolker." *Isis* **44**: pp. 13–16.

—— 1954. "Some Dyes of an Ancient South American Civilization" (trans. of 1953, above). *Dyestuffs* **40**, 9: pp. 238–244.

FESTER, G. A., and J. CRUELLAS. 1934. "Colorantes de Paracas." *Revista del Museo Nacional de Lima* **3**: pp. 154–156.

FESTER, G. A., and S. G. LEXOW. 1941–1943*a*. "Colorantes de insectos." *Anales de la Sociedad Cientifica Argentina* **132**: pp. 100–103; **135**: pp. 89–96.

—— 1943*b*. "Las raices del genero *Relbunium* en la tintoreria americana." *Anales de la Sociedad Cientifica Argentina* **136**: pp. 233–240.

FICKELER, P. 1966. "Fundamental Questions in the Geography of Religions." In: P. L. Wagner and M. W. Mikesell (ed.), *Readings in Cultural Geography* (Chicago), pp. 94–117 (trans. of "Grundfragen der Religionsgeographie," *Erdkunde* **1**, 1947: pp. 121–144).

FIGUIER, L. 1867. *The Vegetable World* (London).

—— 1868. *The Insect World* (London).

FISCHER, E. 1955. "Insektenkost beim Menschen: Ein Beitrag zur Urgeschichte der menschlichen Ernährung und der Bambutiden." *Zeitschrift für Ethnologie* **80**, 1: pp. 1–37.

FLORIO, JOHN. 1598. *A World of Wordes, or Most Copious and Exact Dictionarie in Italian and English* (London).

FORBES, F. B., and W. B. HEMSLEY. 1886–1889. "Index Florae Sinensis: Enumeration of all the plants known from China proper, Formosa, Hainan, Corea . . . the island of Hong Kong." *Journal of the Linnean Society: Botany* **23**.

FORBES, R. J. 1954. "Chemical, Culinary and Cosmetic Arts." In: C. Singer, E. J. Holmyard, and A. R. Hall (ed.), *A History of Technology* (Oxford) **1**: pp. 238–298.

—— 1956. "Dyes and Dyeing." In: *Studies in Ancient Technology* (Leiden) **4**: pp. 98–148.

—— 1964. "Dyes and Dyeing." In: *Studies in Ancient Technology* (2nd ed., Leiden) **4**: pp. 99–150.

FORD, K. C. 1976. *Las hierbas de la gente: a study of Hispano-American medicinal plants* (University of Michigan Museum of Anthropology, paper 60, Ann Arbor).

FOSTER, W. (ed.). 1896–1902. *Letters received by the East India Company from its Servants in the East* (6 v., London) **6** [1617].

—— 1906–1927. *The English Factories in India* (13 v., Oxford) **1** [1618–1621]; **10** [1655–1660].

FRÉZIER, AMADÉE FRANÇOIS. 1717. *A Voyage to the South Sea and along the Coasts of Chili and Peru* [*1712–1714*] (London).

FRIEDBERG, C. 1958. "Contribution à l'étude ethnobotanique des tombes précolombiennes de Lauri, Pérou." *Journal d'Agriculture Tropicale et de Botanique Appliquée* **5**, 6–7: pp. 397–428.

FRIEDERICI, G. 1947. *Amerikanistisches Wörterbuch* (Hamburg).

FRITSCH, K., and J. J. REIN. 1873. "Gegenwärtiger Stand der Cochenille-zucht auf den Canarischen Inseln." *Jahresbericht des Verein für Geographie und Statistik zu Frankfurt au Main* **36**: pp. 49–52.

FUCHSIUS, LEONHARTUS. 1542. *De historia stirpium commentarii insignes* (Basileae).

FURNIVALL, J. S. 1933. *Studies in the Economic and Social Development of the Netherlands East Indies: an Introductory Survey, 1815–1930* (Rangoon).

—— 1939. *Netherlands India* (Cambridge).

GADE, D. W. 1972. "Red Dye from Peruvian bugs." *Geographical Magazine* **45**, 1: pp. 58–62.

—— 1975. *Plants, man and land in the Vilcanota valley of Peru* (Biogeographica 6, The Hague).

GAGE, THOMAS. 1928. *The English-American: a New Survey of the West Indies* [1648] (ed. A. P. Newton, London).

GAMBLE, J. S. 1915–1936. *Flora of the Presidency of Madras* (3 v., London) **1**.

GANDARA, G. 1930. "Flora representada en cerámica Nahoa Pre-Cortesiana." *Anales del Instituto de Biología* (México) **1**, 4: pp. 329–361.

GARCÍA, GREGORIO. 1729. *Origen de los Indios de el Nuevo Mundo, e Indias Occidentales* [1607] (Madrid).

GAY, J. A. 1881. *Historia de Oaxaca* (2 v., México).

GEMELLI CARRERI, JUAN FRANCISCO. 1955. *Viage a la Nueva España* [1700] (ed. and trans. José María de Agreda y Sánchez, 2 v., México).

GENTRY, H. SCOTT. 1972. *Agave Family in Sonora* (United

States Department of Agriculture. Agriculture Handbook no. 399. Washington D.C.).

GEOFFROY, STEPHEN FRANCIS. 1736. *A Treatise of the Fossil, Vegetable and Animal Substances that are made use of in Physick* (trans. G. Douglas, London).

GERARD, JOHN. 1597. *The Herball or Generall Historie of Plantes* (London).

—— 1633. *The Herball or Generall Historie of Plantes* (enlarged and amended by Thomas Johnson, London).

GERARD, J. G. 1833. Letter, dated June 21, 1883, from Herāt, *Journal of the Asiatic Society of Bengal* **2**: p. 652.

GERHARD, P. 1964. "Shellfish Dye in America." *XXXV Congreso Internacional de Americanistas* (México, 1962), *Actas y Memorias* (México) **3**: pp. 177–191.

GERTH VAN WIJK, H. L. 1911. *A Dictionary of Plant Names* (The Hague).

GESNER, KONRAD. 1541. *Historia plantarum et vires ex Dioscoride* (Parisiis).

GIBSON, C. 1952. *Tlaxcala in the Sixteenth Century* (New Haven).

—— 1964. *The Aztecs under Spanish Rule* (Stanford).

GILBERTI, MATURINO. 1901. *Diccionario de la lengua tarasca ó de Michoacán* [1559] (México).

GILL, W. J. 1880. *The River of Golden Sand* (2 v., London).

GILLESPIE, J. E. 1920. *The Influence of Oversea Expansion on England to 1700* (New York).

GILMORE, R. M. 1950. "Fauna and Ethnozoology of South America." In: *Handbook of South American Indians* (ed. J. W. Steward, Smithsonian Institution, Bureau of American Ethnology, Bulletin 143, Washington D.C.) **6**: pp. 345–464.

GIRARD, A. 1932. *Le commerce français à Seville et Cadiz au temps des Hapsbourg* (Paris).

GIRARDIN, J. P. L. 1860–1861. *Leçons de chimie élémentaire appliquée aux arts industriels* (2 v., 4th ed., Paris).

GIRAULT, L. 1966. "Classification vernaculaire des plantes médicinales chez les Callawaya, médecins empirique (Bolivie)." *Journal de la Société des Américanistes de Paris*, nouvelle série, **55**: pp. 155–200.

GOETHE, J. W. VON. 1840. *Theory of Colours* [*Zur Farbenlehre*, 1810] (trans. C. L. Eastlake, London).

GOITEIN, S. D. (trans.). 1973. *Letters of Medieval Jewish Traders.* Translated from the Arabic with introductions and notes (Princeton).

GÓMEZ DE CERVANTES, GONZALO. 1944. *La vida económica y social de Nueva España al finalizar el siglo XVI* [1599] (México).

GONZÁLEZ DE CLAVIJO, RUY. 1928. *Embassy to Tamerlane* [1403–1406] (ed. and trans. G. Le Strange, London).

GONZÁLEZ HOLGUÍN, DIEGO. 1608. *Vocabulario de la lengua general de todo el Peru, llamada Qquichua o del Inca* [1586] (Lima).

GORDON, B. L. 1957. "A Domesticated, Wax-producing, Scale Insect Kept by the Guaymi Indians of Panamá." *Ethnos* (Stockholm) **22**: pp. 36–49.

GRANDIDIER, A. 1917. *Histoire naturelle, physique, et politique de Madagascar* (36 v., 1882–1928, Paris) **4**, 3.

GREEN, E. E. 1912. "On the Cultivated and Wild Forms of Cochineal Insects." *Journal of Economic Biology* 7, 3: pp. 79–93.

—— 1922. *The Coccidae of Ceylon* (London), part 5.

GRIFFITHS, D., and R. F. HARE. 1907. *The Tuna as Food for Man* (United States Department of Agriculture, Bureau of Plant Industry, Bulletin 116, Washington D.C.).

GRISARD, J. 1896. "Emploi des Opuntia dans l'alimentation du bétail." *Bulletin de la société nationale d'acclimatation de France* 43e. ann: pp. 480–485.

GRUBB, W. B. 1904. *Among the Indians of the Paraguayan Chaco* (London).

—— 1911. *An Unknown People in an Unknown Land: An Account of the Lengua Indians of the Paraguayan Chaco* (London).

GUÉRIN-MÉNEVILLE, F. E. 1850. *De la culture de la cochenille en Algérie* (Société national et centrale d'agriculture, Paris).

GUICCIARDINI, LODOVICO. 1593. *The Description of the Low Countreys and of the Provinces thereof, gathered into an Epitome out of the Historie of Lodovico Guicchardini* [1567] by Th. Danett (London).

GUNTHER, R. T. 1922. *Early British Botanists and their Gardens based on Unpublished Writings of Goodyer, Tradescant and others* (Oxford).

HAEFKENS, J. 1969*a*. *Viaje a Guatemala* [*Reize naar Guatemala*, s'Gravenhage, 1827] (Guatemala).

—— 1969*b*. *Centroamerica* [*Centraal Amerika*, Dordrecht, 1832] (Guatemala).

HAENKE, T. 1901. *Descripción del Perú* [*ca.* 1800] (Lima).

HAKLUYT, RICHARD. 1903–1905. *The Principal Navigations, Voyages, Traffiques and Discoveries of the English Nation* (12 v., Glasgow).

—— 1935. *The Original Writings and Correspondence of the two Richard Hakluyts* (ed. E. G. R. Taylor, Hakluyt Society, second series, 2 v., [LXXVI, LXXVII], London).

HALL, FRANCIS. 1824. *Colombia* (London).

ḤAMD-ALLĀH MUSTAUFĪ 1919. *The Geographical Part of the Nuzhatal-Qulūb composed by Ḥamd-Allāh Mustaufī of Qazwīn in 740* [*1340*] (ed. and trans. G. Le Strange, London).

—— 1928. *The Zoological Section of the Nuzhatu-l-Qulūb of Ḥamdullāh Al-Mustaufī Al-Qazwīnī* (ed. and trans. J. Stephenson, London).

HAMEL, J. 1835. "Über Cochenille am Ararat, und über Wurzelcochenille in Allgemeinen." *Memoires de l'académie impériales des sciences de Saint-Pétersbourg,* sixième série, **3**: pp. 9–64.

HAMNET, B. R. 1971*a*. *Politics and Trade in Southern Mexico, 1750–1821* (Cambridge).

—— 1971*b*. "Dye Production, Food Supply and the Laboring Population of Oaxaca." *Hispanic American Historical Review* **51**: pp. 51–78.

HANCE, H. F. 1873. *"Florae Hongkongensis Supplementum." Journal of the Linnean Society: Botany* **13**: pp. 95–114.

HARCOURT, R. D. 1939. *La médecine dans l'ancien Pérou* (Paris).

—— 1960. "Peruvian Textile Techniques." *Ciba Review* **12**, 136: pp. 2–40.

—— 1962. *Textiles of Ancient Peru and their Techniques* [1924] (ed. G. G. Denney and C. M. Osborne, trans. S. Brown, Seattle).

HARCOURT, ROBERT. 1928. *A Relation of a Voyage* [*1608–1609*] *to Guiana by Robert Harcourt* (ed. C. A. Harris, Hakluyt Society, second series, LX, London).

HARDWICK, T. 1822. "Description of a Substance Called Gez or Manna, and the Insect Producing it." *Asiatic Researches* **14**: pp. 182–186.

HARDY, A. 1850. *Culture du nopal: éducation de la cochenille en Algérie* (Alger).

HARDY, R. W. H. 1829. *Travels in the Interior of Mexico, 1825–1828* (London).

HARING, C. H. 1918. *Trade and Navigation between Spain and the Indies in the time of the Habsburgs* (Cambridge).

HAWKS, HENRY. 1904. *A Relation of the Commodities of Nova Hispania* [1572]. In: Richard Hakluyt, *The Principal Navigations, Voyages, Traffiques and Discoveries of the English Nation* (Glasgow) **9**: pp. 378–397.

HAWLEY, W. A. 1913. *Oriental Rugs, Antique and Modern* (New York).

HAYWARD, G. W. 1870. "Journey from Leh to Yarkand and

Kashgar and exploration of the sources of the Yarkana River." *Jour. Royal Geographical Soc.* **60**: pp. 33–166.

HEDDE, I. (J). 1848. *Description méthodique des produits divers recueillis dans un voyage à Chine* (Saint-Étienne).

HEDDE, I. (J), and N. RONDOT. 1849. *Étude pratique du commerce d'exportation de la Chine* (Paris).

HELLOT, J. 1901. *The Art of Dyeing Wool, Silk and Cotton* [1798] (London).

HELMS, A. Z. 1806. *Travels from Buenos Aires by Potosi to Lima* [*Tagenbuch einer Reise durch Peru,* Dresden, 1798] (London).

HEMSLEY, W. B. 1879–1888. *Biologia Centrali-Americana* (4 v., London).

HENRY, A. 1888. "Chinese Names of Plants." *Journal of the China Branch of the Royal Asiatic Society,* n.s. **22**: pp. 233–283.

HERMANN, A. 1969. "Farbe." In: T. Klauser (ed.), *Reallexicon für Antike und Christenstum* (9 v., Stuttgart) **7**: cols. 358–477.

HERNÁNDEZ, FRANCISCO. 1615. [See also Nieremberg, 1635, *infra.*] *Cuatro libros de la naturaleza y virtudes medicinales de las plantas y animales . . . de la Nueva España* [1571–1576], prepared for publication by Francisco Ximénez (México).

—— 1651. *Rerum medicarum Novae Hispaniae thesaurus, seu plantarum, animalium et mineralium Mexicanorum historia* [1571–1576] (Romae).

—— 1959. *Historia natural de Nueva España* (trad. J. Rojo Navarro, 2 v. [of 3 v. of *Obras Completas*] México).

HERNÁNDEZ DE ALBA, G. 1947. "The Highland Tribes of Southern Colombia." In: *Handbook of South American Indians* (ed. J. H. Steward, Smithsonian Institution, Bureau of American Ethnology, Bulletin 143, Washington D.C.) **2**: pp. 915–960.

HERODOTUS. 1963–1969. *Herodotus* (ed. and trans. A. D. Godley, 4 v., Loeb Classics, London).

HERRERA [Y TORDESILLAS], ANTONIO DE. 1725–1726. *General History of the Vast Continent and Islands of America* [*ca.* 1600] (trans. J. Stevens, 6 v., London).

—— 1934–1956. *Historia general de los hechos de los castellanos* [*ca.* 1600] (ed. A. Ballesteros-Beretta, 15 v., Madrid).

HERRERA, F. 1930–1933. *Estudios sobre la flora del departamento del Cuzco* (2 v., Lima).

—— 1934. "Botanica etnológica." *Revista del Museo Nacional* (Lima) **3**: pp. 37–62.

—— 1941. *Sinopsis de la flora del Cuzco* (Lima).

HERRERA, M. 1919. "La cochinilla o pseudococcus cacti." *Revista Agricola* **4**: pp. 164–171.

HEYD, W. 1886. *Histoire du commerce du Levant au Moyen-Age* [*Geschichte des Levantehandels im Mittelalter,* 2 v., Stuttgart, 1879] (2 v., Leipzig).

HILL, A. W. 1915. "The History and Functions of Botanic Gardens." *Annals of the Missouri Botanical Garden* **2**: pp. 185–240.

HILLEBRAND, W. F. 1888. *Flora of the Hawaiian Islands* (London).

HIRE, (M.) DE LA. 1733. Note [on cochineal] (1694). *Histoire de l'Académie Royale des Sciences* (Paris) **2**: pp. 205–206.

HIRTH, F. and W. W. ROOKHILL (ed.) 1911. *Chau Ju-kua: His Work on the Chinese and Arab Trade in the XIIth and XIIIth Centuries Entitled Chu-fan-chi* (St. Petersburg).

HOLMES, W. H. 1889. *Textile Fabrics of Ancient Peru* (Smithsonian Institution, Bureau of American Ethnology, Bulletin 7, Washington D.C.).

HOMER. 1950. *The Iliad* (ed. and trans. E. V. Rieu, London).

—— 1967–1971. *The Iliad* (ed. and trans. A. T. Murray, Loeb Classics, 2 v., London).

HONEGGER, H. VON. 1879. "Einführung und Cultur der Cochenille auf den Kanarischen Inseln." *Der zoologische Garten* **20**: pp. 10–17.

HOOKER, W. J. 1827. "*Cactus cochinillifer:* Spineless Cochineal Fig." *Curtis's Botanical Magazine,* n.s., **1**: pp. 2741–2742.

HOOPER, D. 1910. "Materia Medica Animalium Indica." *Journal and Proceedings of the Asiatic Society of Bengal,* n.s., **6**: pp. 507–522.

HOSIE, A. 1890. *Three Years in Western China* (London).

HUMBOLDT, ALEXANDER VON. 1811. *Political Essay on the Kingdom of New Spain* [1799–1804] (trans. J. Black, 4 v., London).

HUTCHINSON, J., and J. M. DALZIEL. 1927–1928. *Flora of West Tropical Africa* (2 v., London) **1**.

HUTCHINSON, T. J. 1873. *Two Years in Peru, with Exploration of its Antiquities* (2 v., London).

IBN ḤAWQAL. 1800. *Oriental Geography* [*The Routes and the States,* 10th century] (ed. and trans. W. Ouseley, London).

IBN YA'QŪBĪ. 1937. *Ya'ḳūbī: Les Pays* [9th century] (ed. and trans. G. Wiet, Le Caire).

IRVINE, F. R. 1961. *Woody Plants of Ghana* (London).

ITIER, J. 1848–1853. *Journal d'un voyage en Chine en 1843–1846* (3 v., Paris).

J., H. 1938. "Cochineal as Medicine." *Ciba Review* **1, 7**: p. 240.

JACKSON, BENJAMIN DAYDON (ed.). 1876. *A Catalogue of Plants Cultivated in the Garden of John Gerard in the Years 1596–1599* (London).

JACKSON, J. W. 1916. "The Geographical Distribution of the Shell-purple Industry." *Memoirs and Proceedings of the Manchester Literary and Philosophical Society* **60, 7**: pp. 1–29.

JAMESON, B. 1827. "Coccus Cacti." *Edinburgh New Philosophical Journal* **3**: p. 195.

JARAMILLO, ALVARADO, P. 1955. *Historia de Loja y su provincia* (Quito).

JENKINS, K. D. 1964. "*Aje* or *Ni-in* (the Fat of a Scale Insect): Painting Medium and Unguent." *XXXV Congreso Internacional de Americanistas* (México, 1962), *Actas y Memorias* (México) **1**: pp. 625–636.

JENSEN, L. B. 1963. "Royal Purple of Tyre." *Journal of Near Eastern Studies* **22**: pp. 104–118.

JIMÉNEZ DE LA ESPADA, M. (ed.) 1965. *Relaciones Geográficas de Indias* (Biblioteca de Autores Españoles, 4 v. in 3 [CLXXXIII–CLXXXV], Madrid).

JIMÉNEZ MORENO, W. (ed.) 1940. *Codice de Yanhuitlán* [1545–1550] (México).

JINESTA, R. 1940. "Las industrias del añil y de caracol de púrpura." *Revista de los Archivos Nacionales* (San José, Costa Rica) **4**: pp. 302–304.

JOHNSON, I. W. 1954. "Chiptic Cave Textiles from Chiapas, Mexico." *Journal de la Société des Américanistes de Paris,* nouvelle série, **43**: pp. 137–147.

JOHNSON, S. 1755. *A Dictionary of the English Language* (2 v., London).

JONES, C. L. 1966. *Guatemala Past and Present* (New York).

JULIEN, S., and P. CHAMPION. 1869. *Industries anciennes et modernes de l'empire Chinois* (Paris).

JUSTIN-MUELLER, E. 1926. "Cochenille et kermès." *Bulletin de la Société Chimique de France* **39**: pp. 791–792.

KASHA, M. 1948. "Chemical Notes on the Coloring Matter of Chihuahua Textiles in Pre-Columbian Mexico." In: L. M. O'Neale, *Textiles of Pre-Columbian Chihuahua* (Carnegie Institution, Contributions to American Anthropology and History, IX [45], publication 574, Washington D.C.), pp. 151–157.

KATZ, F. 1966. *Situación social y económico de los Aztecas durante los siglos XV y XVI* (México).

KELEMEN, P. 1944–1946. *Medieval American Art* (2 v., New York).

K'ANG-HSI (Emperor). 1779. "Observations de physique et d'histoire naturelle de l'Empereur K'ang-hsi" [*ca.* 1700]. In: *Mémoires concernant l'histoire, les sciences, les arts, les mœurs, les usages . . . des Chinois par les missionaires de Pe-kin* (Paris) **4**: pp. 452–483.

KIRCHHOFF, P. 1948. "Caribbean Lowland Tribes: Mosquito, Sumo, Poya, Jicaque." In: *Handbook of South American Indians* (ed. J. W. Steward, Smithsonian Institution, Bureau of American Ethnology, Bulletin 143, Washington D.C.) **4**: pp. 219–229.

KOUWER, B. J. 1949. *Colours and their Character: a Psychological Study* ('s-Gravenhage).

KRAEMER, H. 1916. "The Nature and Structure of Cochineal." *Journal of the Philadelphia College of Pharmacy* **85**: pp. 344–363.

KURDIAN, H. 1941. "Kirmiz." *Journal of the American Oriental Society* **61**: pp. 105–107 (*cf.* G. L. Della Vida, 1941).

KUROPATKIN, A. N. 1882. *Kashgaria* (trans. W. E. Gowan, Calcutta).

LABAT, J.-B. 1742. *Nouveau voyage* [*1693–1705*] *aux Isles de l'Amerique* (nouv. éd., 8 v., Paris).
and L. Lereboullet (eds.), *Dictionnaire encyclopédique*

LA BOULBÈNE, L. 1876. "Cochenille." In: A. Dechambre and L. Lereboullet (eds.), *Dictionnaire encyclopédique des sciences medicales* (Paris) **18**: pp. 179–198.

LA CRUZ, MARTINUS DE. 1939. *The De La Cruz-Badiano Aztec Herbal of 1552* (ed. and trans. W. Gates, Maya Society, Baltimore).

—— 1940. *The Badianus Manuscript* (*Codex Barberini, Latin 241: Vatican Library*)*: An Aztec Herbal of 1552, trans. by Juannes Badianus into Latin from Aztec* (ed. E. W. Emmart, Baltimore).

LAET, JOHANNE DE. 1628. *Hispania* (Lugduni Batavorum).

—— 1633. *Novus orbis, seu descriptionis Indiae Occidentales* (Lugduni Batavorum).

LANDA, DIEGO DE. 1937. *Yucatan before and after the Conquest* [*ca.* 1566] (ed. and trans. W. Gates, Maya Society, Baltimore).

—— 1941. *Landa's relación de las cosas de Yucatán: a translation* [*ca.* 1566] (ed. and trans. A. M. Tozzer, Papers of the Peabody Museum of American Archeology and Ethnology, Harvard University, **18**).

—— 1966. *Relación de las cosas de Yucatán* [*ca.* 1566] (ed. A. Garibay, México).

LANDÍVAR, RAFAEL. 1948. *Rusticatio Mexicana* [1781], ed. and trans. G. W. Regenos. Middle American Research Institute, Publication 11 (New Orleans), pp. 155–314.

LANGMAN, I. K. 1956. "Botanical Gardens in Ancient Mexico." *Bulletin of the Missouri Botanical Garden* **44**: pp. 17–31.

LAS CASAS, BARTOLEMÉ. 1958. *Apologética Historia* [1559] (ed. J. Pérez de Tudela Bueso, Biblioteca de Autores Españoles, 2 v. [CV, CVI], Madrid).

LATHRAP, D. W. 1966. "Relationships between Mesoamerica and the Andean Areas.". *Handbook of Middle American Indians* (*Archaeological frontiers and external connections*), ed. G. F. Ekholm and G. R. Willey (Austin and London) **4**: pp. 265–276.

LATORRE [Y SETIÉN], G. (ed.). 1919. *Relaciones Geográficas de Indias* (Centro de Estudios Americanistas de Sevilla, Sevilla).

—— 1920. *Relaciones Geográficas de Indias* (Centro de Estudios Americanistas de Sevilla, Biblioteca Colonial Americana, **4**: Sevilla).

LAUFER, B. 1919. *Sino-Iranica* (Chicago).

LAUGHLIN, R. M. 1969. "The Tzotzil." *Handbook of Middle American Indians* (*Ethnology I*), ed. E. Z. Vogt (Austin and London), **7**: pp. 152–194.

LAURENT, J. 1919. *L'Armène entre Byzance et l'Islam* (Paris).

LEE, R. 1948. "Cochineal Production and Trade in New Spain to 1600." *The Americas* **4**: pp. 449–473.

—— 1951. "American Cochineal in European Commerce, 1526–1625." *Journal of Modern History* **23**: pp. 205–224.

LEEWENHOEK, ANTON VAN. 1704. "A Letter from Mr Antony Van Leeuenhoek F. R. S. concerning Cochineal." *Philosophical Transactions of the Royal Society of London* **24**: pp. 1614–1628.

—— 1809. "An Account of Cochineal, 1704." *Philosophical Transactions of the Royal Society of London* (ed. C. Hutton G. Shaw and R. Pearson) **5**: pp. 140–147.

LEGGETT, W. F. 1944. *Ancient and Medieval Dyes* (New York).

LEIX, A. 1937. "Dyes of the Middle Ages." *Ciba Review* **1**, 1: pp. 19–21.

LEMERY, NICOLAS. 1698. *Pharmacopée universelle* (Paris).

LEÓN PINELO, ANTONIO DE. 1943. *El Paraíso en el Mundo Nuevo* [1650] (ed. R. Porras Barrenechea, 2 v., Lima).

LEVILLIER, J. 1928. *Paracas: A Contribution to the Study of Pre-Incaic Textiles in Ancient Peru* (Paris).

LEWIN, B. (ed.). 1958. *Descripción del virreinato del Perú: crónica inédita de comienzos del siglo XVII* (Instituto de Investigaciones Historicas, Universidad Nacional del Litoral, Rosario).

LINCK, J. G. 1787. *Disputatio inauguralis de coccionellae natura, viribus* (Lipsiae).

LINNÉ, SIGVALD. 1948. *El Valle y la Ciudad de México en 1550* (Stockholm).

LINNAEUS, CAROLUS. 1735. *Systema naturae* (Lugduni Batavorum).

—— 1758. *Systema naturae* (10th ed., Holmiae).

—— 1772. "Cochenille de l'arbousier." *Memoires de l'Académie Royale des Sciences de Stockholm* in *Collection Académique* [partie etrangère] (Paris) **11**: pp. 66–67.

—— 1938. *The Critica Botanica of Linnaeus* [1737] (ed. and trans. A. Hort, Ray Society, CXXIV, London).

—— 1957–1959. *Species Plantarum* [facs. of 1st ed. of 1753] (Ray Society, 2 v., [CXL, CXLII], London).

LINOCIER, GEOFROY. 1584. *L'histoire des plantes, traduicte de latin en françois* (Paris).

LINSCHOTEN, JAN HUYGEN VAN. 1598. *Discours of voyages into ye Easte and West Indies* (trans. W. Phillip, London).

LIRA, J. A. 1944. *Diccionario Kkechúwa-Español* (Tucumán).

LISTER, MARTIN. 1672. Letter [concerning cochineal]. *Philosophical Transactions of the Royal Society of London* **7**: p. 5060.

LIZÁRRAGA, REGINALDO DE. 1968. *Descripción breve de toda la tierra del Perú, Tucumán, Río de la Plata y Chile* [*ca.* 1600] (ed. M. Hernández Sánchez-Barba, Biblioteca Autores Españoles, CCXVI, Madrid).

LOBEL, MATTHIAE DE. 1576. *Plantarum seu stirpium historia* [part 2: *Stirpium adversa*, 1570–1571] (Antverpiae).

—— 1591. *Icones stirpium* (2 v., Antverpiae).

LÓPEZ DE GÓMARA, FRANCISCO. 1954. *Historia general de las Indias con la conquista de México y de la Nueva España* [1551–1552] (ed. P. Guibelalde and E. M. Aguilera, 2 v., Barcelona).

—— 1964. *Cortés: The Life of the Conqueror by his Secretary* [1551–1552] (ed. and trans. L. B. Simpson, Berkeley and Los Angeles).

LÓPEZ DE VELASCO, J. 1894. *Geografía y descripción de las Indias* [1571–1574] (ed. J. Zaragoza, Madrid).

LORENZANA, FRANCISCO ANTONIO. 1770. *Historia de Nueva España, escrita por su esclarecido conquistador Hernán Cortés* (México).

LOUREIRO, J. 1790. *Flora Cochinchinensis* (2 v., Ulyssipone).

Löw, I. 1967. *Die Flora der Juden* [repr. of the 1924–1934 ed.] (4 v., Wien and Leipzig).

Loze, A. 1834. *Mémoire sur la culture du nopal et l'éducation de la cochenille* (Alger).

Lund, E. M., and A. C. Tallantire. 1962. *Some Common Flowering Plants of Uganda* (London).

Macgowan (Dr.). 1852–1853. "Abstract of a Paper on the White Wax and the insect by which it is produced." *Proceedings of the Entomological Society of London* 2, 2: pp. 93–95 [from *Journal of the Horticultural and Agricultural Society of India,* 1850].

Mackenzie, A. S. 1831. *A Year in Spain* (2 v., London).

Macleod, M. J. 1973. *Spanish Central America: A Socioeconomic History, 1520–1720* (Berkeley and Los Angeles).

Madoz, Pascual. 1845–1850. *Diccionario geográfico-estadístico-histórico de España y sus posesiones de ultramar* (16 v. in 8, Madrid).

Magán, Vicente. 1963. *Informe* [Santa María Lachixio], *1776.* In: B. Dahlgren de Jordan (ed.), *La grana cochinilla* (México), pp. 73–86.

Magoun, F. P. 1955. "Chaucer's Medieval World outside of Great Britain." *Medieval Studies* **17**: pp. 117–142.

Mahdihassan, S. 1948. "On the Chinese Lac Insect." *Eos* (Madrid) **24**: pp. 441–444.

—— 1950*a*. "The Lac Insect of the West Coast of India." *Journal of the Bombay Natural History Society* **47**: pp. 327–333.

—— 1950*b*. "The Earliest Reference to Lac in Chinese Literature." *Current Science,* September 19, **1950**: pp. 289–290.

—— 1952*a*. "The Mysore Lac Insect." *Journal of the Bombay Natural History Society* **49**: pp. 197–200.

—— 1952*b*. "The Oldest Specimen of Lac in the World." *Indian Journal of Entomology* **14**, 1: pp. 39–43.

—— 1954. "The Natural History of Lac as known to the Chinese: Li Shih-Chen's Contribution to our Knowledge of Lac." *Indian Journal of Entomology* **16**: pp. 309–326.

Mairet, E. 1938. *Vegetable Dyes* [1916] (London).

Mallat, J. 1846. *Les Philippines: histoire, géographie, mœurs, agriculture, industrie et commerce des colonies espagnoles dans l'Océanie* (2 v. in 1, Paris).

Manandyan, H. A. 1965. *The Trade and Cities of Armenia in relation to Ancient World Trade* (trans. N. G. Garsoian, Lisbon).

Markham, C. R. 1892. *A History of Peru* (Chicago).

—— 1908. *Vocabularies of the General Language of the Incas of Peru or Runa Simi* (London).

Martens, E. Von. 1898. "Purpur-Färberei in Central-America." *Zeitschrift für Ethnologie* **30**, 2: pp. 482–486.

Martin, F. R. 1912. *The Miniature Painting and Painters of Persia, India and Turkey from the 8th to the 18th Century* (2 v., London).

Martínez, M. 1939. *Las plantas medicinales de México* (México).

—— 1959. *Plantas utiles de la flora mexicana* (México).

Martius, C. F. P. de. 1843. *Systema materiae medicae vegetabilis Brasiliensis* (Lipsiae Vindobonae).

Matthioli, Pietro Andrea. 1544. *Di P. Dioscoride libri cinque della historia, & materia medicinale tradotti in lingua volgare italiana da P. A. Matthiolo* (Venetia).

—— 1558. *Commentarii secundo aucti, in libros sex Pedacii Dioscoridis Anazarbei de Medica Materia* (Venetiis).

—— 1571. *Compendium de plantis omnibus* (Venetiis).

—— 1586. *De plantis epitome utilissima* (Francofurti ad Moenum).

Maudslay, A. C., and A. P. Maudslay. 1899. *A Glimpse of Guatemala* (London).

Maximilian of Wied-Neuwied, (Prince). 1820. *Travels in Brazil in the years 1815–1817* (London).

McBryde, F. W. 1945. *The Cultural and Historical Geography of South West Guatemala* (Smithsonian Institution, Institute of Social Anthropology, Publication 4, Washington D.C.).

McCulloch, J. R. 1832. *Dictionary of Commerce* (London).

Mead, C. W. 1916. "Conventionalized Figures in Ancient Peruvian Art." *American Museum of Natural History, Anthropological Papers* **12**: pp. 193–225.

Means, P. A. 1932. *A Study of Peruvian Textiles* (Boston).

Megliorini, Juan de, and Santiago de la Cruz. 1826. *Memoria sobre el nopal y cria de la cochinilla de América para el uso de los labradores canarios* (Ciudad de la Laguna).

Meilink-Roelofsz, M. A. 1962. *Asian Trade and European Influence in the Indonesian Archipelago between 1500 and about 1630* (The Hague).

Memorial. 1931. *Copia del Memorial* [concerning cochineal] (1620). In: *Reales Cedulas,* Boletín del Archivo General de la Nación (México) **2**: pp. 497–506.

Mercado, Ignacio de. 1936. *Declaración de las virtudes de los arboles y plantas que estan en este libro* [*ca.* 1695] (ed. D. J. Madrid Moreno, Academia Nacional de Medicina, Madrid).

Merian, Mariae Sibillae. 1719. *Dissertatio de generatione et metamorphosibus insectorum Surinamensium* (Amstelaedami).

Merrifield, Mary P. 1849. *Original Treatises dating from the XIIth to the XVIIIth centuries on the Arts of Painting . . . and the preparation of Colour* (2 v., London).

Merrill, E. D. 1903. *A Dictionary of the Plant Names of the Philippine Islands* (Manila).

—— 1904. *The American Element in the Philippine Flora* (Department of the Interior, Bureau of Government Laboratories, Manila).

—— 1912. *A Flora of Manila* (Manila).

—— 1935. "A Commentary on Loureiro's Flora Cochinchinensis." *Transactions of the American Philosophical Society* **24**, 2.

Métraux, A. 1946. "Ethnography of the Chaco." *Handbook of South American Indians* (ed. J. W. Steward, Smithsonian Institution, Bureau of American Ethnology, Bulletin 143, Washington D.C.) **1**: pp. 197–370.

Meyendorff, G. de. 1826. *Voyage d'Orenbourg à Boukhara, fait en 1820 . . . et revu par M. le Chevalier A. Jaubert* (Paris).

Michell, H. 1955. "Coccus or Kermes." *Classical Review,* n.s., **5**: p. 246.

Middendorf, E. W. 1890. *Wörterbuch des Runa Simi oder der Keshua-Sprache* (Leipzig).

Milburn, W. 1813. *Oriental Commerce* (2 v., London).

Miller, A. G. 1973. *The Mural Painting of Teotiguacán* (Washington D.C.).

Miller, Philip. 1731. *The Gardener's Dictionary* (London).

Ministre de la Guerre, M. le. 1846. Letter [concerning cochineal], with comments by J. B. G. M. Bory de Saint-Vincent. *Comptes Rendus: Académie des Sciences* (Paris) **13**: pp. 220–222.

Minorsky, V. (ed. and trans.). 1937. *Ḥudūd al'Ālam—"The regions of the world": a Persian Geography 372 A.H.–982 A.D.* (London).

Minutes. 1831. *Minutes of Evidence taken before the Select Committee of the House of Lords on the affairs of the East India Company* (Reports from Committees, 5 v., London) **5**, 2.

Minutoli, J. Von. 1854. *Die Canarischen Inseln. Ihre Vergangenheit und Zukunft* (Berlin).

Miquel, F. A. W. 1855–1859. *Flora van nederlandsch Indië* (3 v., Amsterdam).

Miranda, J. 1952. *El tributo indígena en la Nueva España durante el siglo XVI* (México).

MOFFETT, THOMAS. 1658. *The Theater of Insects* [1634] (London).

MOLDENKE, H. N., and A. L. MOLDENKE. 1952. *Plants of the Bible* (Waltham).

MOLINA, ALONSO DE. 1555. *Vocabulario en lengua Castellana y Mexicana* (México).

MOLINA, GIOVANNI IGNAZIO. 1809. *The Geographical, Natural, and Civil History of Chili* [1788–1795] (trans. from the Italian, 2 v., London).

MOLINA SOLIS, J. F. 1896. *Historia del descubrimiento y conquista de Yucatán* (Mérida).

MOLÍNS FÁBREGA, N. 1954. "El códice Mendocino y la economía de Tenochtitlán." *Revista Mexicana de Estudios Antropológicos* **14**: pp. 303–335.

MONARDES, NICOLAS. 1925. *Joyfull Newes out of the Newe Founde Worlde* [*ca.* 1570] (trans. John Frampton, 1577, 2 v., London).

MONOD DE FROIDEVILLE, L. 1847. "Bijdragen tot de kennis van de *Nopal*-kultuur en *Cochenille*-teelt op Java." *Tijdschrift voor nederlandsche Indië* **9**, 2: pp. 207–284.

MONTERO DE MIRANDA, FRANCISCO. 1953. *Descripción de la provincia de la Verapaz* [1574]. Anales de la Sociedad de Geografía e Historia (Guatemala) **27**: pp. 342–358.

MONTGOMERY, G. W. 1839. *Narrative of a Journey to Guatemala in Central America in 1838* (New York).

MOON, A. 1824. *A Catalogue of the Indigenous and Exotic Plants Growing in Ceylon* (Colombo).

MOORCROFT, W., and G. TREBECK. 1841. *Travels in the Himalayan Provinces of Hindustan and the Panjab, in Ladakh and Kashmir, in Peshawar, Kabul, Kunduz, and Bokhara, from 1819–25* (prepared by H. H. Wilson from original journals and correspondence, 2 v., London).

MORELET, A. DE. 1871. *Travels in Central America* [1857] (trans. M. F. Squier, London).

MORENO TOSCANO, A. 1968. *Geografía económica de México* [siglo XVI] (México).

MORENO Y GALINDO, GUILLERMO. 1837. *Instrucción para la cria del insecto americano llamado cochinilla grana en Málaga y su provincial* (Málaga).

MORRIS, D. 1895. "The Plants and Gardens of the Canary Islands." *Journal of the Royal Horticultural Society* **19**, 1: pp. 60–122.

MORRISON, J. R. 1834. *A Chinese Commercial Guide* (Canton).

MORSE, H. B. (ed.). 1926–1929. *The Chronicles of the East India Company Trading to China, 1635–1834* (5 v., Cambridge).

MOTA Y ESCOBAR, ALONSO DE LA. 1945. *Memoriales del Obispo de Tlaxcala* [1608–1624]. Anales del Instituto Nacional de Antropología e Historia de México, **1**: pp. 191–306.

MOTOLINÍA [BENAVENTE] TORIBIO DE. 1941. *Historia de los Indios de la Nueva España* [1541] (ed. D. Sánchez, México).

—— 1951. *History of the Indians of New Spain* [1541] (ed. and trans. Francis Borgia Steck, Academy of American Franciscan History, Washington D.C.).

—— 1970. *Memoriales* [*ca.* 1550] (ed. F. de Lejarza, Biblioteca de Autores Españoles, CCXL, Madrid).

MUMFORD, J. K. 1900. *Oriental Rugs* (New York).

MUÑOZ CAMARGO, DIEGO. 1948. *Historia de Tlaxcala* [*ca.* 1576] (ed. A. Chavero, México).

MURBECK, S. V. 1905. *Contributions à la connaissance de la flore du Nord-Ouest de l'Afrique* (Lund).

MURILLO, G. [Dr. Atl]. 1925. "Tejidos de púrpura de Oaxaca." *Mexican Folkways* **1**, 2: pp. 22–23.

MURRA, J. V. 1962. "Cloth and its Function in the Inca State." *American Anthropologist* **64**: pp. 710–728.

MURÚA, MARTÍN DE. 1962. *Historia general del Perú, origen y descendencia de los Incas* [*ca.* 1600] (ed. M. Ballestros-Gaibros, 2 v., Madrid).

NADER, L. 1969. "The Zapotec of Oaxaca." *Handbook of Middle American Indians (Ethnology,* I), ed. E. Z. Vogt (Austin and London) **7**: pp. 329–359.

NEUGEBAUER, R., and J. ORENDI. 1909. *Handbuch der Orientalischen Teppichkunde* (Leipzig).

NEWBERIE, JOHN. 1625. *Two Voyages* (1578–) of *Master John Newberie.* In: *Purchas his Pilgrimes* (4 v., London) **2**: pp. 1410–1422.

NIEREMBERG, JUAN EUSEBIO. 1635. *Historia Naturae Maxime Peregrinae* (Antwerpt).

NIEUHOFF, J. 1704. *Voyages and Travels into Brasil and the East Indies.* In: A. Churchill (ed. and trans.), *A Collection of Voyages and Travels* (8 v., London) **2**: pp. 1–369.

NOBREGA, G. JOSÉ DE. 1848–1849. "On the Cultivation of Cochineal" (substantially a translation of a "treatise" by Santiago de Cruz e Gonçalves). *Pharmaceutical Journal and Transactions* **8**: pp. 342–348.

NORDENSKIÖLD, E. VON. 1931. "Ancient Inca Lacquer Work." In: *Comparative Ethnological Studies* (Göteborg) **9**: pp. 95–100.

NORIEGA, J. M. 1919. *Las plantas mexicanas y algunas exóticas productores de materias colorantes* (México).

NOWOTNY, K. A., and R. STREBINGER. 1958. "Der Codex Becker I: Technische Beschreibung und Mikroanalytische Untersuchung der Farbstoffe." *Archiv für Volkerkunde* **13**: pp. 222–226.

NOYES, E. (ed. and trans.). 1932. *Fray Alonso Ponce in Yucatán, 1588.* Middle American Research Institute, Publication 4 (New Orleans).

NUTTALL, Z. 1909. "A Curious Survival in Mexico of the use of the Purpura Shell-fish for Dyeing." In: *Anthropological Essays presented to F. Ward Putnam* (Cedar Rapids, Iowa), pp. 368–384.

OBER, F. A. 1887. *Travels in Mexico and Life among the Mexicans* (London).

OBLITAS POBLETE, E. 1969. *Plantas medicinales de Bolivia* (La Paz).

O'CONOR, J. E. 1876. *Lac: Productions, Manufacture and Trade* (Calcutta).

OLIVIER (M.) 1791. "Cochenille." In: *Encyclopédie methodique* (Paris and Liége) **106**: pp. 91–100.

O'NEALE, L. M. 1937. *Textiles of the Early Nazca Period.* Field Museum of Natural History, Anthropology Memoirs (Chicago) **2**: pp. 117–218.

—— 1945. *Textiles of Highland Guatemala* (Carnegie Institution, Publication 567, Washington D.C.).

—— 1949. "Weaving." *Handbook of South American Indians* (ed. J. H. Steward, Smithsonian Institution, Bureau of American Ethnology, Bulletin 143, Washington D.C.) **5**: pp. 97–138.

O'NEALE, L. M., and A. L. KROEBER. 1930. *Textile Periods in Ancient Peru.* University of California Publications in American Archaeology and Ethnology **28** (Berkeley and Los Angeles), pp. 23–56.

OROZCO Y BERRA, M. 1960. *Historia antigua y de la conquista de México* [1880] (ed. A. M. Garibay and M. León-Portilla, 4 v., México).

ORTA, GARCIA DA. 1913. *Colloquies on the Simples and Drugs of India* [1563] (ed. and trans. C. R. Markham, London).

ORTON, J. 1870. *The Andes and the Amazon* (New York).

OSBECK, P. 1771. *A Voyage to China and the East Indies* (trans. J. R. Forster, 2 v., London).

OSBORNE, L. DE J. 1935. *Guatemala Textiles.* Middle American Research Institute, Publication 6 (New Orleans).

—— 1965. *Indian Crafts of Guatemala and El Salvador* (Norman).

OVIEDO Y VALDÉS, GONZALO FERNÁNDEZ DE. 1526. *De la natural hystoria de las Indias* (Toledo).

—— 1534. *Summario de la naturale et general historia de l'Indie occidentali* (n.p.).

—— 1535. *La historia general de las Indias* [primera parte] (Sevilla).

—— 1547. *Cronica de las Indias: La hystoria general de las Indias* [primera parte] (Salamanca [no pagination]).

—— 1555. *L'histoire naturelle en generalle des Indes* (trad. J. Poleur, Paris).

—— 1950. *Sumario de la natural historia de las Indias* (ed. J. Miranda, México and Buenos Aires).

—— 1959*a*. *Natural History of the West Indies* [1526], ed. and trans. S. A. Stoudemire. University of North Carolina, Studies in Romance Languages and Literatures, **32** (Chapel Hill).

—— 1959*b*. *Historia General y Natural de las Indias* [1526–1555] (ed. J. Pérez de Tudela Bueso, Biblioteca de Autores Españoles, 5 v. [CXVII–CXXI], Madrid).

PALACIO (LIC.) 1866. *Relación hecha por el licenciado Palacio al Rey D. Felipe II, en la que describe la provincia de Guatemala* [1576]. Colección de documentos inéditos relativos al descubrimiento, conquista y organización de las antiguas posesiones españoles de América y Oceania (Madrid) **6**: pp. 5–40.

PALMER, E. 1889. "Opuntia Fruit as an Article of Food." *West American Scientist* **6**: pp. 67–69.

PARHAM, J. W. 1964. *Plants of the Fiji Islands* (Suva).

PARKER, R. N. 1918. *A Forest Flora for the Punjab* (Lahore).

PARKINSON, JOHN. 1629. *Paradisi in sole Parodisus Terrestris* (London).

PARRY, J. H. 1966. *The Spanish Seaborne Empire* (London).

PARSON, E. C. 1966. *Mitla: Town of Souls* [1936] (Chicago).

PASO Y TRONCOSO, FRANCISCO (ed.). 1905–1906. *Papeles de la Nueva España* (6 v., Madrid) 1 [*Suma de Visitas*], **4, 5** [*Relaciones Geográficas*].

PATIÑO, V. M. 1963. *Plantas cultivadas y animales domésticos en América equinoccial* (Cali).

PAUSANIUS. 1965–1969. *Description of Greece* (ed. and trans. W. H. S. Jones, 4 v., Loeb Classics, London).

PAZ SOLDAN, D. D. M. 1862. *Geografía del Perú* (Paris).

PEGOLOTTI, FRANCESCO BALDUCCI. 1936. *La Practica della Mercatura* [*ca.* 1310–1340] (ed. A. Evans, Cambridge, Mass.).

PEMEL, ROBERT. 1652–1653. *A Treatise of the Nature and Qualities of such Simples as are most frequently found in Medicines* (London).

PEÑA, R. 1901. *Flora Cruceña* (Sucre).

PEÑAFIEL, A. 1885. *Nombres geográficos de México* (2 v., México).

—— 1890. *Monumentos del arte Mexicano antiguo* (3 v. [2 of plates], Berlin).

PEREIRA, J. 1839–1840. *The Elements of Materia Medica* (2 v., London).

PÉREZ ARBELAEZ, E. 1947. *Plantas utiles de Colombia* (Bogotá).

PÉREZ, V., and P. SAGOT. 1865–1866. De la végétation au Îles Canaries." *Journal de l'Agriculture des Pays Chauds* **1**, 2: pp. 87–99; **3**: pp. 129–148; **4–5**: pp. 223–234; **6–7**: pp. 281–295.

PÉREZ Y LÓPEZ, ANTONIO XAVIER. 1791–1798. *Teatro de la legislación universal de España é Indias* (28 v., Madrid).

PERKIN, A. G., and A. E. EVEREST, 1918. *The Natural Organic Colouring Matters* (New York).

PETRUSHEVSKY, I. P. 1968. "The Socio-economic Conditions of Iran under the Īl-Khāns." In: *The Cambridge History of Iran* (ed. J. A. Boyle, Cambridge) **5**: pp. 483–537.

PFISTER, R. 1928. "La décoration des étoffes d'Antinoë." *Revue des Arts Asiatique* **5**: pp. 215–243.

—— 1934–1940. *Textiles de Palmyre* (3 v., Paris).

—— 1935. "Teinture et alchimie dans l'Orient Hellénistique." *Seminarium Kondakovianum* **7**: pp. 1–59.

Pharmacopoea. 1659. *Pharmacopoea Belgica* (London).

—— 1682. *Pharmacopoea Londinensis* (London).

—— 1771. *Pharmacopoea Helevetica* (Basileae).

—— 1788. *The New Pharmacopoea of the Royal College of Physicians of London* (trans. T. Healde, London).

—— 1825. *Pharmacopoea Rossica et Fennica* (Mediolani).

PHILIPS, MILES. 1904. *A Discourse* [*ca.* 1580]. In: Richard Hakluyt, *The Principal Navigations, Voyages, Traffiques and Discoveries of the English Nation* (Glasgow) **9**: pp. 398–455.

PHIPPS, J. 1835. *A Practical Treatise on the China and Eastern Trade* (Calcutta).

PICKERING, C. 1854–1876. *The Geographical Distribution of Animals and Plants* (2 v., Boston and London).

—— 1879. *Chronological History of Plants* (Boston).

PIKE, R. 1966. *Enterprise and Adventure: The Genoese in Seville and the Opening of the New World* (Ithaca).

PINEDA, E. 1845. *Descripción geográfica del departamento de Chiapas y Soconusco* (México).

PINEDA, JUAN DE. 1925. *Descripción de la provincia de Guatemala: Año 1594*. Anales de Sociedad Geografía e Historia de Guatemala **1**, 4: pp. 327–363.

PISO, GULIEMUS. 1658. *Historiae naturalis et medicae Indiae Occidentalis* (Amstelaedami).

PITARD, J., and L. PROUST. 1909. *Les Îsles Canaries: flore de l'archipel* (Paris).

PLANCHON, G. 1864. *Le kermès du chêne aux points de vue zoologique, commercial et pharmaceutique* (Paris, Montpellier).

PLINY [THE ELDER]. 1961–1968. *Natural History* (ed. and trans. H. Rackham, W. H. S. Jones and D. E. Eichholz, Loeb Classics, 10 v., London).

PLUMIER, C. 1704. "Response du R. P. Charles Plumier Minime à M. Frideric Richter . . . sur la cochenille." *Memoires pour l'histoire des sciences et des beaux arts* (2nd ed., Amsterdam) **8**: pp. 211–222.

—— 1733. Letter [1696]. In: *Histoire de l'Académie Royale des Sciences* (Paris) **2**: p. 280.

POMET, P. 1709. *Droguier curieux: catalogue des drogues, simples et composées* (Paris).

—— 1735. *Histoire générale des drogues* [1694] (2 v., Paris).

PONCE, ALONSO. 1872. *Relación breve y verdadera de algunas cosas de las muchas que sucedieron al padre fray Alonso Ponce en las provincias de Nueva España* [*ca.* 1590] (ed. M. Fernánde de Navarrete, Colección de documentos inéditos para la historia de España, 2 v., [LVII, LVIII], Madrid).

—— 1932. *Fray Alonso Ponce in Yucatán, 1588,* ed. and trans. E. Noyes. Middle American Research Institute, Publication 4 (New Orleans).

PONTING, K. G. 1973. "Logwood: An Interesting Dye." *Journal of European Economic History* **2**, 1: pp. 109–119.

POPE, A. U. (ed.). 1938–1939. *A Survey of Persian Art* (6 v. and index [1958] London).

PORRAS BARRENECHEA, R. (ed.). 1937. *Las relaciones primitivas de la conquista del Perú* (Paris).

POST, G. E. 1932–1933. *Flora of Syria, Palestine and Sinai* (2 v., Beirut).

POSTHUMUS, N. W. 1946. *Inquiry into the History of Prices in Holland* (2 v., Leiden).

POWELL, D. 1972. *The Botanic Garden, Liguanea* (*with a revision of Hortus Eastensis*). Bulletin of the Institute of Jamaica, Science Series **15**, 1.

PRESAS, J. DE. 1825. *Instrucción para el cultivo de la planta nopal . . . y cria de la cochinilla* (Málaga).

PRITCHARD, E. H. 1957. "Private Trade between England and China in the 18th Century (1680–1833)." *Journal of the Economic and Social History of the Orient* **1**: pp. 108–137, 221–256.

QUESADA, V. G. (ed.). 1865. *Noticias sobre la intendencia de Córdoba del Tucumán* [1788]. La Revista de Buenos Aires **6**: pp. 558–605.

RAFFLES, T. S. 1817. *History of Java* (2 v., London).

RAMÍREZ, BALTASAR. 1936. *Description del Reyno del Piru* [1597]. In: H. Trimborn *Quellen zur Kulturgeschichte des präkolumbischen Amerika* (Stuttgart), pp. 1–122.

RAMUSIO, GIOVANNI BATTISTA (ed.). 1563–1565. *Navigationi et viaggi* (3 v., Venetia).

RATZEL, F. 1878. *Aus Mexiko* (Breslau).

RAY, JOHN. 1686–1704. *Historia Plantarum* (3 v., London).

RAYNAL, G. T. F. 1776. *A Philosophical and Political History of the Settlement and Trade of the Europeans in the East and West Indies* [1770] (trans. J. Justamond, 4 v., Dublin).

READ, B. E. 1937. *Chinese Materia Medica: Turtle and Shellfish Drugs* [based on the *Pen T'sao Kang Mu*, 1596] (Peiping).

REAL SOCIEDAD ECONÓMICA DE AMIGOS DEL PAÍS DE CÁDIZ. 1825. *Instrucción para el cultivo del nopal cria de la cochinilla de América* (Cádiz).

RECLUS, É. 1882. *Nouvelle géographie universelle* (19 v., Paris) 7.

REICHEL-DOLMATOFF, G. 1968. *Amazonian Cosmos* (Chicago).

REINA, R. E. 1969. "Eastern Guatemalan Highlands: the Pokomanes and Chorti." *Handbook of Middle American Indians* (*Ethnology*, I), ed. E. Z. Vogt (Austin and London) **7**: pp. 101–132.

REISCHAUER, R. K. 1937. *Early Japanese History, c. 40 BC–AD 1167* (Princeton).

REKO, B. P. 1919. "De los nombres botánicos Aztecas." *El México Antiguo* **1**, 5: pp. 113–157.

Reports. 1839. *Reports on cochineal.* Transactions of the Agricultural and Horticultural Society of India, **6** (Appendix).

REYES, ANTONIO DE LOS. 1750. *Arte en lengua mixteca* [1593] (Puebla).

RICHTER, C. F. 1701. *Tractatus physicus de cochinilla ejusque natura atque indole* (Leipzig).

ROBELO, C. A. 1904. *Diccionario de Aztequismos* (Cuernavaca).

—— 1951. *Diccionario de mitología nahuatl* [1905] (México).

ROBELO, C. A., M. DE OLAGUIBEL, and A. PEÑAFIEL. 1966. *Nombres geográficos indígenas del estado de México* (México).

ROBINSON, B. B. 1947. "Minor Fibre Industries." *Economic Botany* **1**, 1: pp. 47–56.

ROBINSON, ELIZABETH. 1917. "Coccidae of the Philippine Islands." *The Philippine Journal of Science* **12** (D): pp. 1–43.

ROBINSON, V. 1882. *Eastern Carpets* (London).

ROE, THOMAS. 1899. *The Embassy of Sir Thomas Roe to the Court of the Great Mogul, 1615–1619* (ed. W. Foster, Hakluyt Society, second series, 2 v. [I, II], London).

ROJAS GONZÁLEZ, F. 1949. "Los Zapotecas en la epoca prehispanica." In: L. Mendieta y Núñez (ed.), *Los Zapotecos* (México), pp. 35–102.

ROLAND-GOSSELIN, R. 1909. "Notes sur les opuntias cochenillicolles." *Bulletin du muséum national d'histoire naturelle* **15**: pp. 507–509.

ROSSIGNÓN, J. 1859. *Manual del cultivo de añil y del nopal ó sea extracción del indigo, educación y cosecha de la cochinilla* (Paris).

ROTH, W. E. 1915. *An Inquiry into the Animism and Folklore of the Guiana Indians.* Report of the Bureau of American Ethnology for 1908–1909, Smithsonian Institution (Washington D.C.), pp. 103–386.

—— 1924. *An Introductory Study of the Arts, Crafts and Customs of the Guiana Indians.* Report of the Bureau of American Ethnology for 1916–1917, Smithsonian Institution (Washington D.C.), pp. 27–745.

—— 1929. *Additional Studies of the Arts, Crafts and Customs of the Guiana Indians.* Bureau of American Ethnology, Smithsonian Institution, Bulletin 91 (Washington D.C.).

ROWE, C. 1972. "Conceptions of Colour and Colour Symbolism in the Ancient World." *Eranos* **41**: pp. 327–364.

ROXBURGH, W. 1799. "On the La'csha' or Lac insect." *Asiatic Researches* **2**: pp. 361–364.

—— 1820–1832. *Flora Indica* (3 v., London).

ROYLE, J. F. 1840. *Essay on the Productive Resources of India* (London).

ROYS, R. L. 1931. *The Ethnobotany of the Maya.* Middle American Research Institute, Publication 2 (New Orleans).

—— 1933. *The Book of Chilam Balam of Chumayel* (ed. and trans. R. L. Roys. Carnegie Institution, Publication 438, Washington D.C.).

—— 1943. *The Indian background of colonial Yucatán* (Carnegie Institution, Publication 548, Washington D.C.).

RUBIO MAÑE, J. I. (ed.) 1938. In: F. V. Scholes, *Documentos para la historia de Yucatán, III* [1766] (Mérida).

RUBIO SÁNCHEZ, M. 1961. "La grana o cochinilla." *Anales de Antropología e Historia de Guatemala* **13**: pp. 15–46.

RUDDLE, K. 1973. "The Human Use of Insects: Examples from the Yukpa." *Biotropica* **5**, 2: pp. 94–101.

RUELLIUS, JOANNES. 1536. *De natura stirpium libri tres* (Parisiis).

RUIZ, HIPÓLITO. 1952. *Relación histórica del viage, que hizo a los reynos del Perú y Chile el botánico D. Hipólito Ruiz en el año de 1777 hasta el de 1788* (ed. J. Jaramillo-Arango, 2 v., Madrid).

RUIZ DE MONTOYA, ANTONIO. 1876. *Arte de la lengua Guarani, ó mas bien Tupi* [1639] (Paris and Vienna).

RUMPHIUS, GEORGIUS EVERHARDUS. 1741–1750. *Herbarium Amboinense* (6, v., Amsterdam).

RUUSSCHER, MELCHIOR DE. 1729. *Histoire naturelle de la cochenille, justifiée par des documents authentiques* (Amsterdam).

—— 1730. "The Natural History of Cochineal, being an Account of a Book entitled: *Histoire naturelle de la cochenille*" [W. Rutty]. *Philosophical Transaction of the Royal Society of London* **36**: pp. 246–248.

SAGE, C. J. 1907. "Cochineal Cultivation." *The Chemist and Druggist* **81**: p. 140.

SAHAGÚN, BERNARDINO DE. 1938. *Historia general de las cosas de Nueva España* [*ca.* 1570] (ed. J. Ramírez Cabañas, 5 v., México).

—— 1961–1963. *Florentine Codex: General History of the Things of New Spain* [*ca.* 1570] (ed. and trans. C. E. Dibble and A. J. O. Anderson, book X (1961), XI (1963), Santa Fé).

SALTZMAN, M., A. M. KEAY, AND J. CHRISTENSEN. 1963. "The Identification of Colorants in Ancient Textiles." *Dyestuffs* **44**, 8: pp. 241–251.

SÁMANOS, JUAN DE (based on Francisco de Xérez). 1844. *Relación de los primeros descubrimientos de Francisco Pizzaro y Diego de Almagro* [1527]. Colección de documentos inéditos para la historia de España (Madrid) **5**: pp. 193–201.

SAN ANTÓN MUÑÓN, FRANCISCO DE [CHIMALPAHIN CAUHTLEHUANITZIN]. 1965. *Relaciones originales de Chalco Amaquemecan* [*ca.* 1607] (ed. and trans. S. Rendón, México and Buenos Aires).

SANDERSON, JOHN. 1931. *The Travels of John Sanderson in the Levant, 1584–1602* (ed. W. Foster, Hakluyt Society, second series, LXVII, London).

SANTO TOMAS, DOMINGO DE. 1560. *Vocabulario de la lengua general de los Indios del Peru llamada Quichua* (Valladolid).

SANZ ANTÓN, J. 1950. "Un capitulo olivado del ayer económico canario: la cochinilla." *Estudios Geográficos* (Madrid) **11**: pp. 73–91.

SARTORIUS, C. 1961. *Mexico about 1850* (Stuttgart).

SAVARY DES BRUSLONS, J. 1751–1755. *The Universal Dictionary of Trade and Commerce* [1723] (trans. with additions, M. Postlethwayt, 2 v., London).

SAYOUS, A. E. 1929. "Partnerships in Trade between Spain and America and also in the Spanish Colonies in the 16th Century." *Journal of Economic and Business History* **1**: pp. 282–301.

SCHAEFER, G. 1941. "The Cultivation of Madder." *Ciba Review* **4**, 39: pp. 1398–1406.

SCHAFER, E. H. 1957. "Rosewood, Dragon's Blood and Lac." *Journal of the American Oriental Society* **77**: pp. 129–136.

SCHENDEL, G. 1968. *Medicine in Mexico: from Aztec Herbs to Betatrons* (Austin).

SCHMIDT, M. 1929. *Kunst und Kultur von Peru* (Berlin).

SCHMIEDER, O. 1930. *The Settlements of the Tzapotec and Mije Indians, State of Oaxaca, Mexico.* University of California Publications in Geography **4** (Berkeley and Los Angeles).

SCHMITT, C. B. 1971. "Theophrastus in the Middle Ages." *Viator* **2**: pp. 251–270.

SCHOFF, W. H. (ed. and trans.), 1912. *The Periplus of the Erythraean Sea. Travel and Trade in the Indian Ocean by a Merchant of the 1st Century* (London).

SCHOLES, F. V. 1936. "Tributos de los Indios de la Nueva España, 1536." *Boletín del Archivo General de la Nación* **7**, 2: pp. 185–226.

SCHOLES, F. V., and E. B. ADAMS (ed.). 1957. *Información sobre los tributos que los indios pagaban a Moctezuma: Año 1554.* Documentos para la Historia del México Colonial (México) **4**.

SCHOLES, F. V., and E. B. ADAMS (ed.). 1958. *Sobre la manera de tributar los indios de Nueva España a su Magestad, 1561–1564.* Documentos para la Historia del México Colonial (México) **5**.

SCHOLES, F. V., and R. L. ROYS. 1948. *The Maya Chontal Indians of Acalan-Tixchel* (Carnegie Institution, Publication 560, Washington D.C.).

SCHOMBURGK, R. 1848. *Reisen in Britisch-Guiana in den Jahren 1840–1844* (3 v., Leipzig).

SCHURZ, W. L. 1959. *The Manila Galleon* [1939] (New York).

SCHUTZENBERGER, P. *Traité des Matières colorantes* (2 v., Paris).

SCHUYLER, E. 1876. *Turkistan: Notes of a Journey in Russian Turkistan, Khokana, Bukhara and Kuldja* (2 v., London).

SÉBIRE, A. 1899. *Les plantes utiles du Sénégal* (Paris).

SEIDE, J. 1953. "The Relationship of Garcia da Orta's and Cristobal Acosta's Botanical Works." *Actes du septième congrès international d'histoire des sciences* (Jerusalem), pp. 564–567.

SELER, E. 1902–1923. *Gesammelte Abhandlungen zur Amerikanischen Sprach und Alterhumskunde* (5 v., Berlin).

SERMET, J. 1973. "Acclimatation: les jardines botaniques espagnols au XVIIIe. siècle et la tropicalisation de l'Andalousie." In: E. Privat (ed.), *Mélanges en l'honneur de Fernand Braudel—histoire economique du monde. mediterranéen, 1450–1650* (Paris), pp. 555–582.

SEURAT, L.-G. 1930. *Exploration zoologique de l'Algérie de 1830–1930* (Paris).

SHAW, THOMAS. 1738. *Travels or Observations Relating to Several Parts of Barbary and the Levant* (Oxford).

SIGNORET, V. 1875. "Essai sur les cochenilles." *Annales de la Société entomologique de France*, 5e. série, **5**: pp. 17–40, 305–394.

SILLIMAN, BENJAMIN (Jr.). 1871. "The Chinese Wax Insect." *American Naturalist* **5**: pp. 683–685.

SIMPSON, L. B. 1937. "The Medicine of the Conquistadores: An American Pharmacopoea of 1536." *Osiris* **3**: pp. 142–164.

SKINNER, A. 1910. "The Use of Insects and other Invertebrates as Food by the North American Indians." *Journal of the New York Entomological Society* **18**: pp. 264–267.

SLEEMAN, W. H. 1838. "On the Lac Insect and Mode of Cultivating it." *Transactions of the Agricultural and Horticultural Society of India* **6**: pp. 47–51.

SLOANE, HANS. 1696. *Catalogus plantarum quae in insulae Jamaica* (London).

—— 1707–1725. *A Voyage to the Islands of Madera, Barbados, Nieves, S. Christopher's, and Jamaica, with the Natural History of the last of those Islands* (2 v., London).

SMITH, FREDERICK PORTER. 1871. *Contributions toward the Materia Medica and Natural History of China* (Shanghai and London).

SMITH, JOHN. 1882. *A Dictionary of Popular Names of the plants which furnish the natural and acquired wants of man* (London).

SMITH, JAMES EDWARD. 1821. *A Selection of the Correspondence of Linnaeus and other Naturalists, from the Original Manuscripts* (2 v., London).

SMITH, R. S. 1940. *The Spanish Guild Merchant* (Durham, North Carolina).

SOEST, G. H. VAN. 1869–1871. *Geschiedenis van het Kulturrestelsel* (3 v., Rotterdam).

SOLIS Y RIVADENEYRA, ANTONIO DE. 1684. *Historia de la conquista de México* (Madrid).

—— 1753. *The History of the Conquest of Mexico by the Spaniards* (trans. T. Townsend, 2 v., London).

SPALLANZANI, LAZZARO. 1809. *Travels in the Two Sicilies* [late 18th century]. In: J. Pinkerton, *General Collection of the Best and Most Interesting Voyages and Travels in all Parts of the World* (17 v., London) **5**: pp. 1–272.

SPENCE, J. M. 1878. *The Land of Bolivar* [Venezuela] (2 v., London).

SPIX, J. B. VON, and K. F. P. VON MARTIUS. 1824. *Travels in Brazil in the years 1817–1820* (trans. H. E. Lloyd, 2 v. in 1, London).

SPORES, R. 1967. *The Mixtec Kings and their People* (Norman).

SQUIER, E. G. 1852. *Nicaragua* (2 v., New York).

—— 1858. *The States of Central America* (New York).

—— 1870. *Honduras* (London).

START, L. E. 1963. *The McDougall Collection of Indian Textiles from Guatemala and Mexico.* Pitt Rivers Museum, Occasional Papers on Technology, 2 (Oxford).

STATE PAPERS. 1860. *Calendar of State Papers: Colonial Series, 1574–1660* (ed. W. N. Sainsbury, H.M.S.O., London).

—— 1870. *Calendar of State Papers: Colonial Series, East Indies, China and Japan, 1617–1621* (ed. W. N. Sainsbury, H.M.S.O., London).

—— 1884. *Calendar of State Papers: Colonial Series, East Indies and Persia, 1625–1629* (ed. W. N. Sainsbury, H.M.S.O., London).

—— 1892. *Calendar of State Papers: Colonial Series, East Indies and Persia, 1630–1634* (ed. W. N. Sainsbury, H.M.S.O., London).

—— 1894. *Calendar of State Papers and Manuscripts relating to English affairs existing in the Archives and Collections of Venice* (ed. H. F. Brown, H.M.S.O., 38 v., London) **8** [1581–1591].

—— 1899. *Calendar of Letters and State Papers relating to English affairs preserved in, or originally belonging to, the Archives of Simancas* (ed. M. A. S. Hume, H.M. S.O., 4 v., London) **4** [1587–1603].

—— 1904. *Calendar of State Papers and Manuscripts relating to English affairs existing in the Archives and Collections of Venice* (ed. H. F. Brown, H.M.S.O., 38 v., London) **11** [1607–1610].

—— 1916. *Calendar of State Papers relating to English affairs preserved principally at Rome* (ed. J. M. Rigg, H.M. S.O., 2 v., London) **1** [1558–1571].

—— 1928. *Calendar of State Papers: Colonial Series, America and West Indies, 1714–1715* (ed. C. Headlam, H.M. S.O., London).

—— 1964. *List and Analysis of State Papers: Foreign Series, Elizabeth I* (H.M.S.O., 2 v., London) **1** [1589–1590].

Staunton, G. 1797. *An Authentic Account of an Embassy from the King of Great Britain to the Emperor of China* (2 v., London).

Stebbing, E. P. 1907. "A Note on the Lac Insect (*Tachardia lacca*), its Life History, Propagation and Collection." *Indian Forest Records* **1**, 1: pp. 1–84.

Steggerda, M. 1941. *Maya Indians of Yucatán* (Carnegie Institution, Publication 531, Washington D.C.).

Stephens, J. L. 1846. *Incidents of Travel in Central America, Chiapas and Yucatán* [1841] (2 v., New York).

Stoll, O. 1886. *Reisen und Schilderungen aus den Jahren 1878–1883* (Leipzig).

Stone, B. C. 1970. *The Flora of Guam* (Agana).

Stone, D. 1948. "The Northern Highland Tribes: the Lenca [of Honduras]." *Handbook of South American Indians* (ed. J. H. Steward, Smithsonian Institution, Bureau of American Ethnology, Bulletin 143, Washington D.C.) **4**: pp. 205–217.

—— 1949. *The Boruca of Costa Rica.* Papers of the Peabody Museum of American Archaeology and Ethnology, Harvard University, **26**, 2.

Stone, O. M. 1887. *Tenerife and its Six Satellites, or the Canary Islands Past and Present* (2 v., London).

Strabo. 1960–1969. *The Geography of Strabo* (ed. and trans. H. L. Jones, 8 v., Loeb Classics, London).

Stuart, G. A. 1928. *Chinese Materia Medica* (Shanghai).

T. 1937. "Purple Shell-fish on the English Coast." *Ciba Review* **1**, 4: pp. 131–132.

Tamara, Francisco. 1556. *El libro de las costumbres de todas las gentes del mundo y de las Indias, traduzido y copilado por F. Thamara* [from J. Boëmus and C. Giglio] (Anvers).

Tawney, R. H., and E. Power (ed.). 1924. *Tudor Economic Documents* (3 v., London).

Taylor, F. S., and C. Singer. 1956. "Pre-scientific Industrial Chemistry." In: C. Singer *et al.* (ed.), *A History of Technology* (5 v., Oxford) **2**: pp. 347–374.

Taylor, W. B. 1972. *Landlord and Peasant in Colonial Oaxaca* (Palo Alto).

Teixeira, Pedro. 1902. *The Travels of Pedro Teixeira* (ed. and trans. W. F. Sinclair, Hakluyt Society, second series, IX, London).

Termer, F. 1936–1941. *Zur Geographie der Republik Guatemala* (2 v., Geographische Gesellschaft, Hamburg XLIV [1936], XLVII [1941]).

Terry, A. R. 1834. *Travels in the Equatorial Regions of South America in 1832* (Hartford).

Theophrastus. 1961–1968. *Enquiry into Plants* (ed. and trans. A. Hort, 2 v., Loeb Classics, London).

Thiéry de Menonville, N. J. 1787. *Traité de la culture du nopal et de l'éducation de la cochenille dans les colonies françaises de l'Amerique: précédé d'un voyage a Guaxaca* (Cap-Française).

—— 1812. *Travels in Guaxaca in 1777.* In: J. Pinkerton, *General Collection of the Best and Most Interesting Voyages and Travels in all Parts of the World* (17 v., London) **13**: pp. 753–876.

Thompson, G. A. 1829. *Narrative of an Official Visit to Guatemala from Mexico* (London).

Thompson, J. E. S. 1973. *The Civilization of the Mayas* [1927]. Field Museum of Natural History, Anthropology Series **25** (Chicago).

Tilli, Michael Angelo. 1723. *Catalogus plantarum horti Pisani* (Florentiae).

Tillyard, R. J. 1926. *The Insects of Australia and New Zealand* (Sydney).

Tomson, Robert. 1904. *The Voyage of Robert Tomson Marchant, into Nova Hispania in the yeere 1555.* In: Richard Hakluyt, *The Principal Navigations, Voyages, Traffiques and Discoveries of the English Nation* (Glasgow) **9**: pp. 338–358.

Torquemada, Juan de. 1723. *Monarquia Indiana* [1615] (3 v., Madrid).

Tovar, Juan de. 1860. *Historia de los yndios mexicanos* [*ca.* 1600] (ed. T. Phillipps, Middle Hill).

Towle, M. A. 1961. *The Ethnobotany of Pre-Columbian Peru.* Viking Fund Publications in Anthropology, **30** (Chicago).

Tradescant, John. 1656. *Musaeum Tradescantianum: Catalogus plantarum in horto Johannis Tredescanti nascentium* (London).

Tragus, Hieronymus. 1546. *Kreüter Buch* (Strassburgh).

—— 1552. *De stirpium . . .nomenclaturis . . .commentariorum libri III* (Argentorati).

Trémeau de Rochebrune, A. 1879. "Recherches d'ethnographie botanique sur la flore des sépultures péruviennes d'Ancon." *Actes de la Société Linnéene de Bordeaux* **33**: pp. 343–358.

Trigault, Nicolas. 1618. *Histoire de l'expedition chrestienne en la Chine, enterprise par les Pères de la Compagnie de Iesus* [1615] (Paris).

Tristram, H. B. 1898. *The Natural History of the Bible* (9th ed., London).

Tryon, H. 1910. "The 'Wild Cochineal Insect,' with Special Reference to its Injurious Action on Prickly Pear . . ." *Queensland Agricultural Journal* **25**: pp. 188–197.

Tsaï-tin-lang. 1878. *Memoires d'un voyageur chinois dans l'empire d'Annam* [*ca.* 1825]. In: Recueil d'itinéraires et de voyages dans l'Asie Centrale et l'extrême orient. Publications de l'École de Langues orientales vivantes, 7 (trans. from Russian by L. Leger, Paris).

Turner, C. W. 1937. "Colour Symbolism in India." *Ciba Review* **1**, 2: pp. 57–61.

Tylor, E. B. 1861. *Anahuac* (London).

Ulloa, Antonio de, and Jorge Juan Ulloa. 1748. *Relación histórica del viage a la América meridional* [1735–1744] (3 v., Madrid).

—— 1760. *A Voyage to South America* [1735–1744] (trans. J. Adams, 2 v., London).

Uricoechea, E. 1871. *Gramatica, vocabulario, catecismo i confesionario de lengua Chibcha* (Paris).

Vaillant, G. C. 1935. *Artists and Craftsmen in Ancient Central America* (American Museum of Natural History, New York).

Valette, M. 1913. "Note sur la teinture de tissus précolombiniens du Bas-Pérou." *Journal de la Société des Americanistes de Paris,* nouvelle série, **10**: pp. 43–45.

Vambéry, A. 1864. *Travels in Central Asia* (London).

Vargas Machuca, Bernardo de. 1599. *Milicia y descripción de las Yndias* (Madrid).

Vasco, Joaquín. 1963. *Informe* [Santa María Acatepec], *1776.* In: B. Dahlgren de Jordan (ed.), *La grana cochinilla* (México), pp. 45–69.

Vavilov, N. I. 1951. *The Origin, Variation, Immunity and Breeding of Cultivated Plants* [*ca.* 1935] (trans. K. Starr Chester, Chronica Botanica, 13, Waltham).

Vázquez de Espinosa, Antonio. 1942. *Compendio y descripción de las Indias Occidentales* (ed. and trans. [English]

C. U. Clark, Smithsonian Institution, Miscellaneous Collections, CII, Washington D.C.) [text in Spanish, *ibid.*, CVIII, 1948].

VEGA-GRANDE (COMTE DE). 1865. "Note sur la culture de la cochenille aux Canaries." *Bulletin mensuel de la société impériale zoologique d'acclimatation,* 2e. série, **2**: p. 415.

VELASCO, A. L. 1891. *Geografía y estadística del estado de Oaxaca de Juárez.* In: *Geografía y estadística de la República mexicana* (México) **9**.

VETANCURT, AGUSTÍN DE. 1698. *Teatro Mexicano: descripción breve de los sucesos ejemplares, histórico, políticos, militares y religiosos del Nuevo Mundo Occidental de las Indias* (México).

VILLA ROJAS, A. 1969*a*. "Maya Lowlands: the Chontal, Chol, and Kekchi." *Handbook of Middle American Indians* (*Ethnology I*) ed. E. Z. Vogt (Austin and London) **7**: pp. 230–243.

—— 1969*b*. "The Maya of Yucatán." *Handbook of Middle American Indians* (*Ethnology I*), ed. E. Z. Vogt (Austin and London) **7**: pp. 244–275.

VILLASEÑOR Y SÁNCHEZ, JOSÉ ANTONIO DE. 1746–1748. *Theatro Americano: descripción general de los reynos y provincias de la Nueva España y sus jurisdicciones* (2 v., México).

VILLAVICENCIO, M. 1858. *Geografía de la República del Ecuador* (New York).

VITRUVIUS. 1962. *On Architecture* (ed. and trans. F. Grainger, Loeb Classics, 2 v., London).

WAEIJ, H. W. VAN. 1859. "Bijdrage tot kennis der kultuur van eenige handelsprodukten van Java." *Tijdschrift voor nederlandsche Indië* **21**, 1: pp. 35–46.

WAFER, LIONEL. 1934. *A New Voyage and Description of the Isthmus of America* [1699] (ed. L. E. Elliott Joyce, Hakluyt Society, second series, LXXIII, London).

WARMINGTON, E. H. 1928. *The Commerce between the Roman Empire and India* (Cambridge).

WATT, G. 1889–1893. *Dictionary of the Economic Products of India* (6 v. in 9, Calcutta).

—— 1901. "Lac (*lakh*) and Lac Industries." *Agricultural Ledger* **8**, 9: pp. 181–347.

—— 1908. *The Commercial Products of India* (London).

WATT, J. M., and M. G. BREYER-BRANDWIJK. 1932. *The Medicinal and Poisonous Plants of Southern Africa* (Edinburgh).

WAVRIN (MARQUIS DE). 1937. *Mœurs et coutumes des Indiens sauvages de l'Amérique du Sud* (Paris).

WEBER (DR.) 1896. "Agaves et Cactées." *Bulletin de la société nationale d'acclimatation de France,* 43e. ann: pp. 257–262.

—— 1900. "Le figuier de Barbarie." *Bulletin de la société nationale d'acclimatation de France,* 47e. ann: pp. 5–12.

WEITLANDER, R. J. 1969. "The Cuicatec." *Handbook of Middle American Indians* (*Ethnology I*) ed. E. Z. Vogt (Austin and London) **7**: pp. 434–447.

WEST, R. C. 1948. *Cultural Geography of the Modern Tarascan Area* (Smithsonian Institution, Institute of Social Anthropology, Publication 7, Washington D.C.).

WEST, R. C., and J. P. AUGELLI. 1966. *Middle America: Its Lands and Peoples* (Englewood Cliffs).

WESTHEIM, P. 1948. "Colour in Mexican Life." *Ciba Review* **6**, 70: pp. 2576–2580.

WHITNEY, GEFFREY. 1586. *A Choice of Emblemes* (Leyden).

WIEPEN, E. 1890. *Die geographische Verbreitung der Cochenillezucht* (Bonn).

WIGHT, R. 1840–1850. *Illustrations of Indian Botany* (2 v., Madras).

WIGHT, R., and G. A. WALKER-ARNOTT. 1834. *Prodromus Florae Peninsulae Indiae Orientalis* (London) **1** (no more published).

WILLETTS, W. 1958. *Chinese Art* (2 v., London).

WILLIAM, R. O. 1949. *The Useful and Ornamental Plants in Zanzibar and Pemba* (Zanzibar).

WILSON, E. H. 1913. *A Naturalist in Western China* (2 v., London).

WILSON, F. 1960. *A Review of the Biological Control of Insects and Weeds in Australia and Australian New Guinea* (Commonwealth Agricultural Bureaux, Farnham Royal).

WISDOM, C. 1940. *The Chorti Indians of Guatemala* (Chicago).

WITTMACK, L. 1890. "Die Nutzplanzen der alten Peruaner." *VII Congrès International de Américanistes* (Berlin, 1888), *Compte-Rendu* (Berlin), pp. 325–349.

XÉREZ, FRANCISCO DE. 1853. *Verdadera relación de la conquista del Perú y provincia del Cuzco* [1543] (ed. E. de Vedia, Biblioteca de Autores Españoles, XXVI, Madrid), pp. 319–343.

—— 1872. *Narrative of the Conquest of Peru* [1543] (ed. and trans. C. R. Markham, Hakluyt Society, XLVII, London), pp. 1–109.

XIMÉNEZ, FRANCISCO. 1957. *Historia natural del reino de Guatemala* [1722] (Guatemala).

YACOULEFF, E., and F. HERRERA. 1934–1935. "El mundo vegetal de los antiguos peruanos." *Revista del Museo Nacional* (Lima) **3**: pp. 241–322; **4**: pp. 29–102.

YACOULEFF, E., and J. C. MUELLE. 1934. "Notes al trabajo *Colorantes de Paracas.*" *Revista del Museo Nacional* (Lima) **3**: pp. 157–163.

YEATS, J. 1887. *The Natural History of the Raw Materials of Commerce* (3rd ed., London).

ZAHAN, D. 1972. "White, Red and Black: Colour Symbolism in Black Africa." *Eranos* **41**: pp. 365–396.

ZÁRATE, BARTOLOMÉ DE. 1915. *Relación descriptiva del valle de Oaxaca* [1544]. In: M. Cuevas (ed.), *Cartas y otros documentos de Hernán Cortés* (Sevilla), pp. 253–256.

ZORITA, ALONSO DE. 1864. *Breve y sumaria relación de los señores, y maneras y diferencias que habia de ellos en la Nueva España.* In: *Colección de documentos inéditos relativos de descubrimiento, conquista y colonización de las posesiones Españoles en America y Oceánia* (Madrid) **2**: pp. 1–126.

—— 1909. *Historia de la Nueva España* [*ca.* 1575] (Madrid).

INDEX